GENETICS - RESEARCH AND ISSUES

HUMAN GENOME

COMPONENTS, STRUCTURAL/FUNCTIONAL DISORDERS AND ETHICAL ISSUES

GENETICS - RESEARCH AND ISSUES

Additional books in this series can be found on Nova's website under the Series tab.

Additional e-books in this series can be found on Nova's website under the e-book tab.

GENETICS - RESEARCH AND ISSUES

HUMAN GENOME

COMPONENTS, STRUCTURAL/FUNCTIONAL DISORDERS AND ETHICAL ISSUES

TOMEO CACCAVELLI
EDITOR

New York

Library of Congress Cataloging-in-Publication Data

ISBN: 978-1-62808-803-8

Library of Congress Control Number: 2013946509

Published by Nova Science Publishers, Inc. † *New York*

Contents

Preface   vii

Chapter I   Cytogenetics to Genomics
                   in Cancer Diagnostics   1
                   G. Sharma, R. Tapadia, A. Rao
                   and M. R. Kollabattula

Chapter II   Evolution of Human Genome Analysis:
                   Impact on Diseases Diagnosis and
                   Molecular Diagnostic Labs   143
                   Julie Gauthier, Isabelle Thiffault,
                   Virginie Dormoy-Raclet
                   and Guy A. Rouleau

Index   173

Preface

The human genome is the complete set of human genetic information, stored as DNA sequences within the 23 chromosome pairs of the cell nucleus and in a small DNA molecule within the mitochondrion. In this publication, the authors present topical research in the study of the components, structural and functional disorders; and ethical issues of the human genome. Topics discussed include critical events in the pathobiology of cancer and the compilation of major biomarkers enabling improved diagnosis; and the evolution of human genome analysis and its impact on disease diagnosis and molecular diagnostics.

Chapter I – Cancer is a major health problem worldwide and an early effective diagnosis would ensure timely management, impacting on the quality of life, the patients overall well being and longevity. There is no single test that can accurately diagnose cancer. An effective diagnostic test should be able to confirm or eliminate the presence of disease, monitor the disease process, and plan for and evaluate the effectiveness of treatment. Though there is a wide array of methods to diagnose cancer, an effective biomarker is still an unmet medical need.

Conventional tests use single genes or discrete pathways. Cancer is a consequence of multiple changes occurring both in parallel and in sequence in the macromolecules of the cell. Earlier only gross end points were possible to be identified and measured, but today with technological advances it has become possible to interrogate events that are very early and at scales hitherto irresolvable. Thus changes in these cellular events are studied at different levels.

DNA changes in cancer are expressed in terms of gross chromosomal anomalies or could be in the sequence of the gene or gene product/protein. As researchers learn more about the mechanisms of cancer, new diagnostic tools are constantly being developed and existing methods refined.

Diagnostic procedures for cancer may include imaging, laboratory tests (including tests for tumor markers), tumor biopsy, endoscopic examination, surgery or genetic testing. This chapter draws attention to the critical events in pathobiology of cancer as is resolved by today's technological progress while compiling major biomarkers enabling improved diagnosis and reviewing some putative biomarkers for translation from bench to bedside.

Chapter II – The advent of massively parallel sequencing has changed the interrogation process of the human genome and now provides a high resolution and global view of the genome which is beyond research applications. Together with powerful bioinformatics tools, these next generation sequencing technologies have revolutionized fundamental research and have important consequences for clinically actionable tests, diagnosis and treatment of rare diseases and cancers. Today, molecular testing is commonly used to confirm clinical diagnosis of specific diseases; it requires that a clinician specify the gene or mutation to test and, in return, will receive information only about this sequence. Despite relative successes, a large number of patients receive no accurate diagnosis, even after many expensive molecular investigations. A clear paradigm shift has taken place in the health network with the introduction of the exome sequencing in molecular diagnostic lab. In this chapter, the impact of the implementation of high throughput sequencing technologies on molecular diagnosis and on the practice of medicine, with an emphasis in paediatrics, is reviewed. We compared well-established genetic tests, using examples from our molecular diagnostic lab, to the recent exome sequencing applications. The genetic tests can fall into three main categories: 1) Mendelian Single Gene Disorder tests that include targeted mutation and targeted gene approaches 2) Genetic Disease Panels which are composed of a few to a dozen genes and 3) Exome or Genome approaches, which interrogate either the entire coding sequences of the 22,333 human genes or the entire human genome. For each of these categories, advantages and limitations are discussed. We devoted a section on the future of molecular diagnosis and discuss which tests will subsist and which one may be soon abandonned. Massively parallel sequencing is transforming the molecular diagnostic field: it offers personalized genetic tests and generates new ethical challenges. Important questions like incidental findings and possible forms of discrimination are addressed. Finally, we conclude with a section on the future directions surrounding the application of these multimodal molecular approaches in general and their putative applications in neonatal intensive care units.

In: Human Genome
Editor: Tomeo Caccavelli

ISBN: 978-1-62808-803-8
© 2013 Nova Science Publishers, Inc.

Chapter I

Cytogenetics to Genomics in Cancer Diagnostics

G. Sharma, R. Tapadia, A. Rao and M. R. Kollabattula
Tapadia Diagnostic Centre Pvt Ltd
RTC X Roads, Hyderabad, India

Abstract

Cancer is a major health problem worldwide and an early effective diagnosis would ensure timely management, impacting on the quality of life, the patients overall well being and longevity. There is no single test that can accurately diagnose cancer. An effective diagnostic test should be able to confirm or eliminate the presence of disease, monitor the disease process, and plan for and evaluate the effectiveness of treatment. Though there is a wide array of methods to diagnose cancer, an effective biomarker is still an unmet medical need.

Conventional tests use single genes or discrete pathways. Cancer is a consequence of multiple changes occurring both in parallel and in sequence in the macromolecules of the cell. Earlier only gross end points were possible to be identified and measured, but today with technological advances it has become possible to interrogate events that are very early and at scales hitherto irresolvable. Thus changes in these cellular events are studied at different levels.

DNA changes in cancer are expressed in terms of gross chromosomal anomalies or could be in the sequence of the gene or gene product/protein. As researchers learn more about the mechanisms of cancer, new diagnostic tools are constantly being developed and existing methods refined.

Diagnostic procedures for cancer may include imaging, laboratory tests (including tests for tumor markers), tumor biopsy, endoscopic examination, surgery or genetic testing. This chapter draws attention to the critical events in pathobiology of cancer as is resolved by today's technological progress while compiling major biomarkers enabling improved diagnosis and reviewing some putative biomarkers for translation from bench to bedside.

Introduction

The central dogma of biology is that information flow is: from DNA to RNA to Protein to function. The protein could be as translated or it could be modified by glycosylation, acytelation etc. The DNA in eukaryotes has a higher order of organisation called the chromosome wherein chromosome is shortened by more than 100,000 fold by super imposing different levels of coiling. At the core of the structure, DNA is wrapped around a cluster of histone proteins which is an octamer made of histones (2 dimers of H2A and H2B, and an H3/H4 tetramer). The fundamental DNA- histone complex thus formed is called the nucleosome. About 147 base pairs of DNA coil around 1 octamer, and ~20 base pairs are sequestered by the addition of the linker histone (H1), and various length of "linker" DNA (0-100 bp) separate the nucleosomes. Packaging of DNA is facilitated by the electrostatic charge distribution. Phosphate groups cause DNA to have a negative charge, whilst the histones are positively charged molecules as they contain lysine and arginine in larger quantities. So they make a strong ionic bond in between them to form nucleosome. The DNA in the native state exists in the A and B form; however it takes the alternate Z form that is relaxed and enables transcription.

Differences in gene activity can be recognised by the degree of compactness, which is indicated by staining intensity. Densely staining heterochromatin represents non active material. This is normally found at centromeric and telomeric regions. These are non transcribable regions and are known to contain satellite DNA.

Variation in biological material is brought about by two fundamental mechanisms: mutation and recombination. Mutations can occur at two basic levels: 1. Gene mutations and Chromosome mutations. Mutations are sudden changes in the nucleotide sequence which can occur in both somatic and germinal tissues. Once mutation has occurred and if it is not lethal such cells multiply to form a clone, at times both types (the mutant and original) of cells coexist leading to clonal heterogeneity or chimerism. Also mutations are a random process that can occur in any cell at any time. Functionally these mutations could result in five types of alterations: hyper-morph, hypo-morph, neo- morph, anti-morph, and amorp, resulting in over expression, under expression, expression of a novel protein instead a protein that is antagonistic to the original, and one that results in an inactive protein respectively. These mutations could also be differentiated as syntax alterations in the DNA sequence such as gains, deletions, inversions all resulting in reading frame changes. Phenotypically these changes in DNA would result in concomitant changes in information, transcription and translation.

In the context of a disease this may be presented as gene mutations leading to altered gene expression resulting in altered protein status and differential metabolism which in turn is expressed as:

- uncontrolled proliferation
- altered metabolism
- increased vascularisation
- down regulated death

Mutations in cancer cells could occur in any one or more proteins in the metabolic pathway of the cell, bringing in the differentiation between two types of mutations based on their impact: Driver mutations versus Passenger mutations.

Driver mutations are defined as being ones that confer oncogenic properties to the cell such as growth advantage, tissue invasion, metastasis and angiogenesis. Passenger mutations are ones that do not critically contribute to any cancerous property but are just there. Chromosome mutations unlike gene mutations are visible changes and are normally referred to as chromosomal anomalies or aberrations. Chromosomes of any given species/cell are topologically ordered and this provides a way of identifying them. Useful markers in this regard are its size, centromere position, nucleolar organizer position, the chromomere, heterochromatin and banding pattern when subjected to different staining protocols. Mutations or aberrations are again of two types one numerical and the other structural. Numerical anomalies could result in aneuploid or euploid changes. In the earlier type the change could be in one or more chromosome resulting in N+/- 1and in the latter it would impact the whole set of chromosomes resulting in N+/-N. Generally euploid cell do not survive due to gross addition /deletion of information leading to a total imbalance in homeostasis. In terms of structural abnormalities there are four types: deletions, duplications, inversions and translocations. Each of these has a varied impact based on the area as well as the function it encompasses. Large aberrations are discernible by routine cytogenetics/staining however if the region involved is very small it cannot be readily observable. Technologies today, as discussed later in this chapter can detect these. Apart from the chromosomal DNA it is also known that even humans have extra chromosomal DNA, and this is present as mitochondrial DNA or genes. The impact of these genes in cancer is also discussed. Recombination is again an important mechanism resulting in variation. This occurs during meiosis when two homologous chromosomes come together and there is crossover between the pair of chromosomes. This again is a consequence of breakage and reunion, resulting in two daughter chromosomes which now have exchanged alleles at the site of recombination, while other genes remain in the original linkage association. The process of recombination, i. e. breakage and reunion has to be exact or else this leads to anomalous chromosomes with duplication, deletion, inversion or translocations. Both mutation and recombination are subjected to repair and at the end the fate of the cell is subject to all that has happened. Each and every step here is mediated through a signal (internal or external) and a group of enzymes. Thus cell division normal or malignant is initiated by some kind of signal that propels a cell to replicate its DNA and go through a step wise process till two daughter cells result. Thus typically eukaryotic cells, have to complete an ordered series of events called the 'cell cycle' to proliferate which include the faithful replication of their genome and the correct segregation of the two copies generated into two daughter cells at the end of cell division. A disruption of these events may lead to cell death or oncogenic transformation. The processes of cell cycle, therefore, are carefully regulated. A key step in the eukaryotic cell cycle is the G1 to S phase transition and this step is tightly coupled to the transcriptional control of genes involved in growth, DNA replication, repair and cytokinesis. Key regulators of cell cycle play a central role in tumor development as well. Cellular mechanisms exist to ensure an adequate supply of gene and gene products to the cell maintaining a structural, functional, temporal regulation of the genes. Any variation in this results in disease or death. One major disease state recognised as malignant transformation is a consequence of perturbation of chromosome, gene structure, number and regulation. This chapter attempts to draw attention to various findings that have enabled understanding of cancer biology as a fall out of recent technologies as well as due to the discovery of the human genome and the ability to use some of this information to develop diagnostic and prognostic biomarkers. It is hoped that the development of such biomarkers would result in early effective diagnosis, disease management methods.

Background

According to WHO, cancer is responsible for 12 % of all deaths in the world. One in every four deaths in the USA is from cancer. The annual burden of cancer to the US economy amounts to US $ 171. 6 billion as estimated by the National Institute of Health [1]. More than 11 million people are diagnosed with cancer every year and it is estimated that by 2020 there will be 16 million cases [2]. Cancer is thus one of the leading causes of morbidity and mortality where cells undergo uncontrolled proliferation and invade, erode and destroy normal tissue. Driving force behind cancer is damaged genes. Environmental factors such as smoking, radiation and viral infection play a significant role in causing this damage. There is increasing evidence to suggest that cancer is also driven by epigenetic changes like DNA methylation, histone modifications resulting in chromatin condensation status thereby regulating expression of specific genes. There are more than 200 different types of cancer. All have different causes, symptoms, and require different types of treatment . Cancer develop due to interaction between genes, environment and chance. Tumorigenesis is a multistep process involving accumulation of defective cancer genes. The alteration of these multiple genes confers a survival advantage to cells. Cancers may be classified by their primary site of origin or by their histological or tissue types. The international standard for the classification and nomenclature of histologies is the International Classification of Diseases for Oncology [3].

Classification by Site of Origin

By primary site of origin, cancers may be of specific types like breast cancer, lung cancer, prostate cancer, liver cancer, renal cell carcinoma, kidney cancer, oral cancer and brain cancer etc.

Classification by Tissue Types

Based on tissue types cancers may be classified into six major categories:

1. Carcinoma

This type of cancer originates from the epithelial layer of cells that form the lining of external parts of the body or the internal linings of organs within the body. Carcinomas are malignancies of epithelial tissue, and account for 80 to 90 percent of all cancer cases since epithelial tissues are most abundantly found in the body from being present in the skin to the covering and lining of organs and internal passageways, such as the gastrointestinal tract. Carcinomas usually affect organs or glands capable of secretion including breast, lungs, bladder, colon and prostate. Carcinomas are of two types – adenocarcinomas and squamous cell carcinoma. Adenocarcinoma develops in an organ or gland and squamous cell carcinoma originates in squamous epithelium. Adenocarcinomas may affect mucus membranes and are first seen as a thickened plaque-like white mucosa. These are rapidly spreading cancers.

2. Sarcoma

These cancers originate in connective and supportive tissues including muscles, bones, cartilage and fat. Bone cancer is one of the sarcomas termed osteosarcoma. It is found commonly in the young. Sarcomas appear like the tissue in which they have originated. Examples include chondrosarcoma (of the cartilage), leiomyosarcoma (smooth muscles), rhabdomyosarcoma (skeletal muscles), Mesothelial sarcoma or mesothelioma (membranous lining of body cavities), Fibrosarcoma (fibrous tissue), Angiosarcoma or hemangioendothelioma (blood vessels), Liposarcoma (adipose or fatty tissue), Glioma or astrocytoma (neurogenic connective tissue found in the brain), Myxosarcoma (primitive embryonic connective tissue) and Mesenchymous or mixed mesodermal tumor (mixed connective tissue types).

3. Myeloma

These originate in the plasma cells of bone marrow. Plasma cells are capable of producing various antibodies in response to infections. Myeloma is a type of blood cancer.

4. Leukemia

This is a group of cancers that are grouped within blood cancers. These cancers affect the bone marrow which is the site for blood cell production. When cancerous, the bone marrow begins to produce excessive immature white blood cells that fail to perform their usual actions and the patient is often prone to infection.

Types of leukemia include:

- Acute myelocytic leukemia (AML) – these are malignancy of the myeloid and granulocytic white blood cell series seen in childhood.
- Chronic myelocytic leukemia (CML) – this is seen in adulthood.
- Acute Lymphatic, lymphocytic, or lymphoblastic leukemia (ALL) – these are malignancy of the lymphoid and lymphocytic blood cell series seen in childhood and young adults.
- Chronic lymphatic, lymphocytic, or lymphoblastic leukemia (CLL) – this is seen in the elderly.
- Polycythemia vera or erythremia – this is cancer of various blood cell products with a predominance of red blood cells.

5.   Lymphoma

These are cancers of the lymphatic system. Unlike the leukemias, which affect the blood and are called "liquid cancers", lymphomas are "solid cancers". These may affect lymph nodes at specific sites like stomach, brain, intestines etc. These lymphomas are referred to as extra nodal lymphomas.

Lymphomas may be of two types – Hodgkin's lymphoma and Non-Hodgkin's lymphomas. In Hodgkin lymphoma there is characteristic presence of Reed-Sternberg cells in the tissue samples which are absent in Non – Hodgkin lymphoma.

6.   Mixed types

These have two or more components of the cancer. Some of the examples include mixed mesodermal tumor, carcinosarcoma, adenosquamous carcinoma and teratocarcinoma. Blastomas are another type that involves embryonic tissues.

Classification by Grade

Cancers can also be classified according to grade. The abnormality of the cells with respect to surrounding normal tissues determines the grade of the cancer. Increasing abnormality increases the grade, from 1–4. Cells that are well differentiated closely resemble normal specialized cells and belong to low grade tumors. Cells that are undifferentiated are highly abnormal with respect to surrounding tissues. These are high grade tumors.

Grade 1 – well differentiated cells with slight abnormality
Grade 2 – cells are moderately differentiated and slightly more abnormal
Grade 3 – cells are poorly differentiated and very abnormal
Grade 4 – cells are immature, primitive and undifferentiated

Classification by Stage

Cancers are also classified individually according to their stage. There are several types of staging methods. The most commonly used method uses classification in terms of tumor size (T), the degree of regional spread or node involvement (N), and distant metastasis (M). This is called the TNM staging. For example, T0 signifies no evidence of tumor, T1 to 4 signifies increasing tumor size and involvement and T signifies carcinoma in situ or limited to surface cells. Similarly N0 signifies no nodal involvement and N 1 to 4 signifies increasing degrees of lymph node involvement. Nx signifies that node involvement cannot be assessed. Metastasis is further classified into two –M0 signifies no evidence of distant spread while M1 signifies evidence of distant spread. Stages may be divided according to the TNM staging classification. Stage 0 indicates cancer being in situ or limited to surface cells while stage I indicates cancer being limited to the tissue of origin. Stage II indicates limited local spread; Stage III indicates extensive local and regional spread while stage IV is advanced cancer with distant spread and metastasis.

Diagnostic Techniques in Cancer

Cancer Diagnostic Techniques in Clinical Practice Today

There are several methods of diagnosing cancer. With advances in technologies that understand cancers better, there is a rise in number of diagnostic tools that can help detect cancers. Once suspected, diagnosis is usually made by pathologists and onco-pathologists and imaging radiologists and today a specialist in genetics is added to the team.

Some types of cancer, particularly lymphomas, can be hard to classify, even for an expert. Most cancers need a second opinion regarding diagnosis before being sure of the diagnosis or stage and type. Also more than one technique or diagnostic test is carried out before treatment is initiated.

In todays context for better value and health of the patient personalized management is adopted. The unravelling of the human genome, the technological advances that have been made has enabled an increased understanding of the disease, its diagnosis and treatment as well.

The amount of information added necessitates the need to periodically review the information/data added and utilised. This chapter addresses the modern molecular biology based techniques used both in the clinical set up and at the research labs as well; however for reasons of completeness these routine techniques are also mentioned (Figure 1)

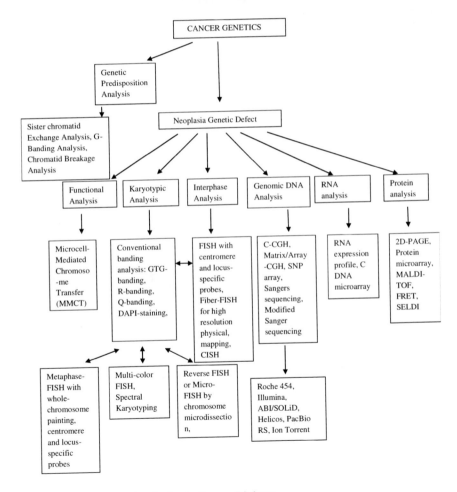

Figure 1. Enabling Technologies in Cancer Biology.

Sampling of the patient: Depending on the test to be carried out appropriate samples are collected, stored, transported and processed. Types of samples include: blood, saliva, urine, faeces, body fluids/exudates, bone marrow aspirates, lymph aspirates and tissue biopsies.

Biopsy: This is a test where a small sample of tissue is taken from the suspected cancer with the help of a fine tipped needle (fine needle aspiration – FNA), or with a thicker hollow needle (core biopsy) or by surgical excision. The tissues are then processed, stained and examined under a microscope for the presence of cancer cells. Depending on tumor location, some biopsies can be done on an outpatient basis with only local anaesthesia and others through surgery

Sentinel node biopsy: This is a procedure where the closest and most important nodes near the cancer are surgically excised and examined.

Pap Test

Pap test (Pap smear) is a routine test where a sample of cells from a woman's cervix is taken, stained and cytology examined under the microscope. This helps identify changes in the cells that could indicate cervical cancer or other conditions.

Sputum Analysis and Bronchial Washing Analysis

The cells of the sputum and bronchial secretions are processed, stained and analyzed under the microscope for signs of lung and other respiratory cancers which are likely to spread; only lymph nodes are likely to contain cancer cells.

The most common diagnostic methods include:

- Histopathology: The morphologic changes identified here refer to the structural alterations in cells or tissues that are either characteristic of the disease or diagnostic of the etiologic process. The practice of diagnostic pathology is devoted to identifying the nature and progression of disease by studying morphologic changes in tissues of patients [4]. The histopathology is done after a biopsy of tissue in query is taken, processed, stained, slide prepared and screened by the pathologist who then reports the finding as per standard guidelines.

- Immunohistochemistry: In many cases, malignant tumors of diverse origin resemble each other because of poor differentiation. These tumors are often quite difficult to distinguish on the basis of routine hematoxylin and eosin-stained tissue sections. The availability of specific monoclonal antibodies has greatly facilitated the identification of cell products or surface markers. Antibodies against intermediate filaments have proved to be of value in such cases. In cases in which the origin of the tumor is obscure, immune-histochemical detection of tissue-specific or organ specific antigens in a biopsy specimen of the metastatic deposit can lead to the identification of the tumor source. Immunohistochemistry in conjugation with immmunofluorescence has also proved useful in the identification and classification of tumors [4].

Endoscopy

In this imaging technique a thin, flexible tube with a tiny camera at the end is inserted into the body cavities. This allows the doctors to view the suspicious area. There are many types of scopes, each designed to view particular areas of the body. For example, a colonoscope looks at the colon and large intestine and a laparoscope is used to look within the abdomen etc.

Blood Tests

Blood tests can be performed to detect the normal blood cells as well as for specific tumor markers. Some tumors release substances called tumor markers, which can be detected in the blood. A blood test for prostate cancer determines the amount of prostate specific antigen (PSA). Higher than normal PSA levels can indicate cancer. Similarly in ovarian cancer a tumor marker CA-125 is released.

Imaging Studies

There are several imaging techniques. These include X rays, CT scans, MRI scans of various parts of the body. *X-rays* are the most common imaging techniques and they may be made more specific by using a Barious enema.

This is used for detection of stomach and small intestinal growths and cancers. *Mammogram* is an X-ray of the breasts used to screen for and/or detect breast lumps and growths. *A CAT scan* (computerized axial tomography) uses radiographic beams to create detailed computerized pictures. It is more precise than a standard X-ray. *Magnetic Resonance Imaging (MRI)* uses a powerful magnetic field to create detailed computer images of the body's soft tissue, large blood vessels and major organs. Both CT scan and MRI can also be used with contrast radio-labelled dyes to obtain a more clear and specific picture of the cancer.

An ultrasound uses high-frequency sound waves to determine if a suspicious lump is solid or fluid. These sound waves are transmitted into the body and converted into a computerized image. Bone scan is specifically used to identify and locate new areas of cancer spread to the bone. Normally a Positron imaging test (PET scan) is used. A Gallium scan is another nuclear medicine test in which a special camera takes pictures of tissues of the body after a special radioactive tracer is injected into a vein. The cancerous areas light up under the scanner.

Genetic Analysis in Cancer Diagnostics

Chromosomes were first observed in plant cells by Karl Wilhemvon Nageli in 1842. In science books, the number of human chromosomes remained at 48 for over thirty years. New techniques were needed to correct this error. Joe Hin Tjio working in albert Levans's lab was responsible for finding the approach by:

1. Using cells in culture
2. Pre-treating cells in a hypotonic solution, which swells them and spreads the chromosomes
3. Arresting mitosis in metaphase by a solution of colchicine
4. Squashing the preparation on the slide forcing the chromosomes into a single plane
5. Cutting up a photomicrograph and arranging the result into an indisputable karyogram.

This initial technology has been modified since then. Today automation at many steps have been incorporated, though fundamentally the principle has remained the same.

Cells from bone marrow, blood, amniotic fluid, cord blood, tumor, and tissues (including skin, umbilical cord, chorionic villi, liver, and many other organs) can be cultured using standard cell culture techniques in order to increase their number. A mitotic inhibitor (colchicines or colcemid) is then added to the culture. This stops cell division at mitosis which allows an increased yield of mitotic cells for analysis. The cells are then centrifuged and media, mitotic inhibitor are removed, and replaced with a hypotonic solution. This causes the white blood cells or fibroblasts to swell so that the chromosomes will spread when added to a slide as well as lyse the red blood cells. The cells have been allowed to incubate in hypotonic solution; Carnoy's fixative (3:1 methanol to acetic acid) is added. This kills the cells and hardens the nuclei of the remaining white blood cells. The cells are generally fixed repeatedly to remove any debris or remaining red blood cells. The cell suspension is then dropped onto specimen slides. After aging the slides in an oven or leaving them for a few days they are ready for banding and analysis.

Chromosomal Analysis

In 1956, it became generally accepted that the karyotype of man included only 46 chromosomes. In 1960, Peter Nowell and David Hungerford discovered the first chromosomal abnormality associated with cancer using cytogenetics (Nowell & Hungerford, 1960). They reported a small chromosome in the lymphocytes of patients with chronic myelogenous leukemia (CML). This has been called Philadelphia chromosome as they were working in Philadelphia, Pennsylvania. In 1973, Janet Rowley used new cytogenetic techniques, namely quinacrine fluorescence and G-banding, to examine patients' karyotypes and demonstrated the translocation of chromosome 9 and 22 which is today an important diagnostic for CML.

Karyotyping

Karyotyping refers to analysis of metaphase chromosomes which have been banded using trypsin followed by Geimsa, Leishman, or a mixture of the two. This creates unique banding patterns on the chromosomes. Several chromosome-banding techniques are used in cytogenetic laboratories. Quinacrine banding (Q-banding) was the first staining method used to produce specific banding patterns. This method requires a fluorescence microscope [5] and is no longer widely used as Geimsa banding (G-banding) [6-8].

Reverse banding, or R-banding, requires heat treatment and reverses the usual black-and-white pattern that is seen in G-bands and Q-bands. This method is particularly helpful for staining the distal ends of chromosomes. Other staining techniques include C-banding [9] and nucleolar organizing region stains (NOR stains) [10]. These latter methods specifically stain certain portions of the chromosome. C-banding stains the constitutive heterochromatin, which usually lies near the centromere, and NOR staining highlights the satellites and stalks of macrocentric chromosomes.

High-resolution banding involves the staining of chromosomes during prophase or early metaphase (prometaphase), before they reach maximal condensation. Because chromosomes in these stages are more extended than metaphase chromosomes, the number of bands observed for all chromosomes increases from about 300 to 450 to as many as 800. This allows the detection of less obvious abnormalities usually not seen with conventional banding.

Analysis

Analysis of banded chromosomes is done with the help of a microscope by a clinical laboratory specialist in cytogenetics (CLSp (CG)). Generally 20 cells are analyzed which is enough to rule out mosaicism to an acceptable level.

The results are summarized and given to a board-certified cytogeneticist for review, and to write an interpretation taking into account the patients previous history and other clinical findings. The results are then given out reported in an International System for Human Cytogenetic Nomenclature 2009 (ISCN2009).

G-banding is still regarded as being the gold standard for genetic tests, since it is the best one currently available for assessing the whole karyotype at once, however subtle or sub microscopic (i. e., cryptic) rearrangements affecting regions smaller than a chromosomal band are extremely difficult to detect by G-banding [11-13].

In situ hybridization: In situ hybridization is a powerful tool in oncology allowing one to analyze the constitution of genomes in a very direct manner. Using preparations of selected nucleic acid sequences as hybridization probes in cellular preparations, this method provides a link between the fields of molecular genetics and classical cytogenetics. Since the first report of the routine isotopic method in 1969 by Gall and Pardue [14] using tritium labelled probes [15-17], the in situ hybridization methods have undergone extensive advancement with regards to both the target and the probe [18, 19].

Fluorescent labelling based techniques replaced the costly and potentially dangerous radioactive techniques used in research for the detection of genetic alterations in tumor cells [20, 21].

Probe labelling techniques have evolved from the original radiographic labelling to fluorescent and chromogenic-based detection, as in fluorescence [22] and chromogenic [23] in situ hybridization, FISH and CISH, respectively. An important extension of conventional FISH methods is the development of multi fluorochrome techniques, such as multiplex FISH (M-FISH) [24], spectral karyotyping (SKY) [25] and combined binary ratio labelling (COBRA) [26], which allow the simultaneous visualization of all chromosomes in 24 colors.

Fluorescence in Situ Hybridisation (FISH)

In addition to standard metaphase preparations FISH can also be performed on:

- Bone marrow smears
- Blood smears
- paraffin embedded tissue preparations
- enzymatically dissociated tissue samples
- uncultured bone marrow
- uncultured amniocytes
- cytospin preparations

The slide is casted and then aged using a salt solution usually consisting of 2X SSC (salt, sodium citrate). The slides are then dehydrated in ethanol, and the probe mixture is added. The sample DNA and the probe DNA are then co-denatured using a heated plate and allowed to re-anneal for at least 4 hours. The slides are then washed to remove excess unbound probe, and counterstained with 4', 6-Diamidino-2-phenylindole (DAPI) or propidium iodide. Analysis of FISH specimens is done by fluorescence microscopy. For oncology generally a large number of interphase cells are scored in order to rule out low-level residual disease, generally between 200 and 1,000 cells are counted and scored. For congenital problems usually 20 metaphase cells are scored.

Currently, the most commonly used conventional in situ hybridization protocol in cancer research is dual-color FISH. FISH analysis should be used as a supplement to conventional cytogenetics.

Dual-color FISH is used for the detection of chromosomal gains or losses (aneuploidy); intra chromosomal insertions, deletions, inversions, amplifications; and chromosomal translocations in both solid and hematopoietic cancers [27-32]. Three different types of probes are commonly used, each with different ranges of applications. They are:

1. Gene specific probes: Gene-specific probes target DNA sequences present in only one copy per chromosome. They are used to identify chromosomal translocations, inversions, duplications and deletions, contiguous gene syndromes, marker chromosomes, and chromosomal amplifications in interphase and metaphase chromosomes [33, 34].
2. Repetitive sequence probes: Repetitive sequence probes bind to chromosomal regions that are represented by short repetitive base-pair sequences that are present in multiple copies (e.g. centromeric and telomeric probes). Centromeric probes are extremely useful for identifying marker chromosomes and for detecting copy number chromosome abnormalities in interphase nuclei. On the contrary, subtelomeric probes are frequently used to identify minute obscure chromosomal translocations [35].
3. Whole-chromosome painting (WCP) probes: FISH with a whole-chromosome painting probe is applicable to metaphase chromosomes and is most helpful in verifying the involvement of specific chromosomes in a translocation . Whole chromosome- painting probes, can be applied to metaphase spreads but not to interphase nuclei [e. g. multicolor FISH (M-FISH) and SKY] [36, 37].

FISH can be used to supplement G banding but is restricted to the defined chromosome regions of the FISH probes used. FISH assay targeting a specific area can aid in providing accurate diagnosis for these indeterminate pathological variants which have considerable clinical and histological overlap [38-41]. However, the greater ability of FISH than cytology to detect early and peripheral disease, can have an impact on overall survival of the patient [42, 43]. In conclusion FISH can detect cells that have chromosomal abnormalities consistent with neoplasia in exfoliative and aspiration cytology specimens. The main limitation of FISH in a clinical setting is standardizing the quantification and interpretation of the fluorescent signal intensities, which typically are averages, based on the analysis of 50-150 interphase or metaphase cells per sample. Thus, selection of individual microscopic areas to be analyzed can be prone to have an observer bias. FISH requires a fluorescence microscope, and the signals are labile and rapidly fade over time. These limitations can be overcome by CISH, (Chromogenic in situ hybridization) which can visualize the amplification product along with morphological features [44, 45].

Interphase FISH: This technique has been used on uncultured amniocytes and chorionic villi as a tool for providing rapid prenatal diagnosis. This eliminates the need for long-term culturing procedures and also provides accurate diagnosis within 24 hours [46-48]. Interphase FISH is a 'direct' approach where single locus probes (SLPs) are used to probe cell nuclei to assess the gross numerical and structural characteristics of a tumor cell population.

High-Resolution Fibre-FISH: The map resolution of fibre-FISH is around 10 kb, which is much higher than metaphase and interphase FISH. High resolution fibre-FISH has been applied to high-resolution physical mapping in the study of DNA-protein in situ interactions to elucidate the functional aspects of chromosome and genome. Fibre-FISH has been used mainly in research rather than clinical laboratories because of the complexity of the procedures involved. [49-56]

Spectral FISH: (S-FISH) is a molecular cytogenetic technique which can target several specific chromosomal aberrations in interphase and metaphase cells in a single hybridization reaction, using a combination of fluorescence and digital imaging microscopy. Spectral FISH can also identify disease-specific aberrations at the DNA level in individual tumor cells. This technique also is used in a clinical setting [57].

3D-FISH: FISH on three-dimensional preserved nuclei (3D-FISH) in combination with three dimensional- microscopy and image reconstruction can analyze entire chromosomes, chromosomal sub regions, or single gene loci on a single-cell level [58]. This however is as of now used as a research tool.

Multicolor-FISH: Conventional FISH allows for the simultaneous analysis of only a few probes and requires a clue for the selection of the appropriate probe(s). In contrast, Multicolor-FISH or M-FISH [24] and spectral karyotyping [25], introduced by Speicher et al. and Schrock et al., respectively, in 1996, allow for the simultaneous identification of all 24 human chromosomes in a different color by a single hybridization with a probe cocktail. To obtain a differential color for each chromosome, the DNA from each of the 24 different flow-sorted chromosomes are isolated, amplified, and labelled with a combination of five fluorescence dyes that can be visualized in different colors through a specific optical image capture and computer analyzing system.

The advantage of M-FISH is it is strong in detecting complex rearrangements of chromosomes and in identification of marker or derivative chromosomes. The multicolor karyotyping approach is very powerful in identifying subtle, complex interchromosomal rearrangements and marker chromosomes which cannot be identified by conventional banding analysis [59-62]. However, hybridization quality and number of metaphases will be the most limiting factor [63]. Multicolor karyotyping, however, cannot detect intra-chromosomal aberrations, such as duplication, deletion, and inversion. It can detect the chromosomes involved, but not the specific breakpoints or chromosomal segments. Recently it was reported that multicolor classification of structural rearrangements between non homologous chromosomes frequently result in overlapping fluorescence at the interface of the translocated segments to generate a "flaring" phenomenon. This flaring can obscure or distort the fluorescence pattern of adjacent chromatin and lead to misinterpretation by the multicolor karyotyping system. To overcome this problem, it is mandatory to confirm the result by DAPI/G banding/ conventional FISH analysis of the specific chromosomes involved. [64, 65]

Reverse FISH by Chromosome Micro dissection: The chromosome micro dissection approach provides a straightforward method for identifying any chromosomal segment of unknown origin by dissecting the segment of interest and isolating, amplifying, fluorescence labelling, and reverse in situ hybridizing DNA to normal metaphase spreads. Once the components of the aberrant chromosome are identified, conventional FISH with whole-chromosome- painting probes can be applied to define the relative position of the components in the marker chromosome. This micro-FISH approach has been applied successfully in the identification of ring chromosomes, homogeneously staining regions and double minutes. Thus, micro-FISH can be used to identify not only the origin of the marker chromosome but also the regions and breakpoints. The skills, procedures, and setup involved in micro-FISH are too complicated to be established in the routine clinical cytogenetic laboratory [37, 66-77].

CISH: CISH is an emerging alternative detection technique using light microscopy and allows a concurrent analysis of histological features of the tumors along with the chromosomal aberrations like gene amplifications and deletions [44, 45]. The first description of the CISH procedure as a practical alternative to FISH for the detection of genetic alterations was published in

2000 [78]. Chromogenic visualization (colorimetric method) in CISH is based on enzyme-conjugated antibodies that recognize the target of interest. Reaction of substrates with enzymes such as horse radish peroxidise (HRP) and/or alkaline phosphatase (AP) leads to chromogen precipitates, which then can be detected with a bright-field microscope. Advantages of this method are

1. CISH results are easily interpreted by the use of a bright-field microscope which is generally used in diagnostic laboratories.
2. CISH enables visualization of the nucleus and is also able to distinguish invasive from in situ carcinomas.
3. CISH signals do not generally fade over time allowing the tissue samples to be archived and reviewed later.
4. CISH resembles immunohistochemistry to a large extent (as opposed to FISH) due the use of conventional counterstains, e. g. hematoxylin, for visualization of tissue morphology.

Chromosomal aberrations identified by different banding techniques mentioned above in more than 47,000 human neoplasms have been catalogued [79]. An increasing number of the acquired abnormalities have now also been studied by various fluorescence in situ hybridization techniques, which have provided a new and powerful tool to identify abnormal chromosomes and to visualize very small rearrangements that escape detection by conventional chromosome banding. The new techniques have also added a further sophistication to the analyses in that breakpoints in structural aberrations can be delineated within specific genes. Furthermore, an ever increasing number of breakpoints of the cancer-associated chromosome abnormalities have been characterized at the molecular level. The combined efforts of cytogeneticists and molecular geneticists over the past two decades have led to the identification of 275 genes rearranged as a consequence of chromosome aberrations in neoplasia. Although targeted FISH is a rapid and highly sensitive technique, the use of specific probes does not exclude other chromosomal abnormalities that may or may not be associated with a particular disorder or malignancy. Thus, it cannot be used for genome-wide screening of chromosomal aberrations for which conventional cytogenetics remains the "gold standard. " [80-83]

Comparative Genomic Hybridisation (CGH): CGH is a molecular cytogenetic method for detecting relative differences in copy number between two genomes. It requires no prior knowledge of chromosomal imbalances; hence, CGH can be used as a discovery tool [18]. It can detect alterations as small as 36 kb [84]. CGH derived data has made clear that chromosomal abnormalities that occur in early premalignant stages can also be detected [85-90]. To perform CGH analysis on a neoplasm, genomic DNA from both tumor and control cells are extracted, differentially labelled with fluorescence dyes, and hybridized in equal amounts to normal reference metaphase spreads. Along the metaphase chromosomes, the color ratio between the two distinct fluorescence colors, which reflect the ratio of the tumor DNA to control DNA, hybridized to the chromosomal regions, are measured with a digital image and a specific software analysis system. The gain (amplification) and loss (deletion) of specific chromosome regions can be detected by increased and decreased ratios of tumor-specific fluorescence color to control [91]. The advantage of the CGH approach is that it can provide genome-wide screening of chromosomal deletion or amplification, that is, copy number changes wherein the steps of culturing cells and obtaining metaphase spreads in the specimens tested is not required. Furthermore, it can be applied to both fresh and fixed and paraffin-embedded material [92]. The major limitations of this technique however, are the inability to detect balanced chromosome rearrangements such as translocations, inversions and clonal heterogeneity [93]. The resolution of chromosome CGH is limited to approximately 10 million base pairs (Mbp). Conventional CGH is labor intensive, providing relatively low resolutions of 5–10 Mb for deletions due to tightly condensed metaphase and 2 Mb for amplifications moreover, it is unsuitable for the detection of balanced rearrangements (e. g. , balanced translocations and inversions), as well as whole genome copy number changes (ploidy) [94-99].

Array CGH: In order to overcome the low resolution limitation of CGH, array CGH (aCGH) was developed. In aCGH, the differentially labelled test and reference DNA is hybridized to arrayed DNA probes on a glass slide rather than metaphase [100-102]. The resolution of aCGH depends on the density and sizes of DNA probes on the array. Although a number of probe substrates have been used in aCGH to date, including large-insert clones (40–200 kb), small insert clones (1.5–2.5 kb), cDNA clones (0.5–2 kb), PCR products (0. 1–1.5 kb), and oligonucleotides (25–85 bp),large insert clones and more recently oligonucleotides have been the most popular(98). Recently, an array CGH has been presented carrying 2400 FISH- and radiation hybrid mapped BAC-clones, covering the whole human genome with an average spacing of 0. 8–1.4 Mbp [103]. Array CGH does not require karyotyping, the resolution can be 0.5 Mbp or better and it has a higher sensitivity than conventional CGH [100, 101]. Rather than using metaphase spreads, an array of very small spots of genomic DNA (BAC clones, representing different genes or chromosome locations) is used as the target for hybridisation. This technical breakthrough provides a locus-by-locus measure of DNA copy-number changes that significantly overcomes some of the limitations of conventional CGH. Low-copy-number gains and losses can be detected by matrix CGH at a resolution about 100 kb, whereas that for high-level amplification can be detected at several kilobases [96]. It has the ability to detect aneuploidy, deletions, duplications, or amplifications of any locus represented on the array [104]. This technique does not require the stimulation of cell cultures by phytohemagglutination (PHA), as classically done in clinical cytogenetics, which may distort the percentage of mosaic cells and inhibit the detection of some mosaic abnormalities by chromosome analysis [105, 106]. Despite their increasing popularity, the main technological limitation of these methods is the restricted applicability to the detection of genome rearrangements that involve a change in copy numbers.

Microcell-Mediated Chromosome Transfer: The cytogenetic detection of non random chromosomal/monosomy or deletion and molecular detection of high frequency of loss of heterozygosity indicate the possible presence of a tumor suppressor gene(s) on the particular chromosome region(s) for a specific tumor type. This indication needs to be verified functionally by either microcell- mediated chromosome transfer at the chromosome level or transfection study at the gene level. To perform microcell-mediated chromosome transfer, a murine cell line carrying a single human chromosome (of interest), tagged with a selectable marker, is used as donor cells for the chromosome transfer. Microcells containing a single chromosome are generated by treating the donor cells with colcemid and cytochalasin-B, collected by centrifugation, and enriched by sequential filtration through polycarbonated membrane with a pore-size decreasing from 8 mm to 3 mm. These microcells are treated with phytohemagglutinin and polyethylene glycol to enhance attachment and fusion, respectively, to the recipient human neoplastic cells. Recipient cells with the human chromosome of exogenous origin are selected in G-418 medium and characterized molecularly and cytogenetically. [107-109]. Correlation between the transfer of a specific chromosome/segment and the restoration of a cellular function for the recipient cell indicates that the function-related gene(s) is located on that particular chromosome/segment. Using this functional approach, the chromosome/regions that bear the gene(s) associated with cellular senescence, tumorigenicity, metastasis progression, and telomerase suppression have been identified. This also is a discovery/research tool. [110-120]

Oligonucleotide arrays (aCGH): This technique is normally performed using arrays of long-oligos designed to hybridize with specific genomic loci. Array CGH allows detection of copy number differences between a test and reference sample of DNA. Oligonucleotide probes are particularly advantageous, as they can be standardized across all arrays used, are devoid of repetitive sequences, and are subsequently much more reproducible. They can be spaced more densely across specific parts of the genome, allowing for better detection of smaller genomic changes and providing increased sensitivity. They can also be customized as information about the genome is updated, and multiple probes can target a single region, allowing for more robust data analysis and increasing reproducibility, sensitivity and confidence in CNV (copy number variations) calls [121]. In contrast, oligonucleotide arrays can provide a higher resolution (generally 5–50 kb) but have been reported to suffer from lower sensitivity. This can result in failure to reliably detect low-copy number changes due to a poorer signal to noise ratio. Oligonucleotide aCGH can potentially provide even higher resolution than 5 kb because overlapping nucleotides can be synthesized with as little as a single base off-set. Although such high resolution cannot currently be achieved on a genome-wide level in a cost-effective manner, it is practical for the study of CNVs in specific regions. [98, 122]

The advantage of aCGH has been adopted by many clinical genetics laboratories as a first-line test for congenital abnormalities due to its automatability and potential to transform cancer genetics due to its detection-resolution for CNVs, deletions, amplifications, duplications and aneuploidies. The disadvantage however is that, small sequence alterations or single base pair mutations will still not be detected; neither will balance chromosomal translocations or inversions. Another disadvantage of standard aCGH is its relative inability to detect areas of LOH (Loss of Heterozygosity) [123].

Representational oligonucleotide microarray analysis (ROMA): Another microarray-based technique, representational oligonucleotide microarray analysis (ROMA), has been developed to address the low signal to noise ratio problem of oligonucleotide aCGH. ROMA uses restriction digestion of genomic DNA followed by the selection of shorter fragments that are then amplified by PCR (Polymerase Chain Reaction) [124]. The principle of ROMA is based on the concept of genomic representation that is generated by amplifying restriction enzyme (such as Bgl II) digested genomic fragments from samples. The representative genomic fragments hybridize to the oligonucleotide in arrays that are designed from the human genome sequence assembly [124, 125].

DNA obtained from reference and test samples is differentially labelled and hybridized to an oligonucleotide array as in aCGH. In this manner, the complexity of the genomic sequence is reduced, because only representative fragments are analyzed by hybridization, thus increasing the signal to noise ratio. ROMA achieves an average resolution of 30 kb, which is comparable with that of oligonucleotide aCGH.

The major advantage of using a representation strategy is to minimize the genome complexity and therefore maximize the signal-to-background ratio. Using ROMA, investigators are able to detect regions of copy number variations between cancer and normal genomes and between normal human genomes . Currently, ROMA can reach a resolution up to 30 to 35 kb [124, 126]. However, several challenges have prevented this method from becoming widely adopted.

1. First, the signal to noise ratio is still lower than that of BAC arrays, and typically, signals from several probes are averaged to reduce noise induced variance in the data.

2. Second, the complexity reduction may lead to unequal representation of different parts of the genome, potentially leading to erroneous CNV calls.

3. Third, restriction digestion patterns may vary among different individuals as a result of restriction fragment length polymorphisms, which ROMA may misinterpret as CNVs (98). Further; the PCR amplification step has the potential of introducing additional biases.

SNP array: SNP arrays, originally designed for genotyping, are oligonucleotide arrays that detect the two different alleles of biallelic SNPs. Probe signal intensities can be used to determine SNP genotypes and to detect copy number changes. In contrast to aCGH, in which samples are differentially labelled and cohybridized, only one labelled sample is hybridized to the SNP array at a time; CNVs are detected by comparison with one or several reference samples analyzed in separate hybridizations. Currently, SNP arrays capable of genotyping more than 900,000 SNPs are available from companies such as Illumina and Affymetrix, providing a resolution that matches or exceeds that of most state-of-the-art aCGH platforms. An important advantage of SNP arrays is the unique ability, among genomic methods discussed thus far, to detect copy number neutral losses of heterozygosity.

Further, SNP arrays have been used to detect allele-specific CNVs, and single nucleotide change. SNP-arrays have been widely applied to the analysis of tumor genomes, including those of ovarian cancer, prostate cancer, colorectal cancer, malignant melanoma and pancreatic cancer. Its limitations is that it requires costly reagents, specialized equipment and the analysis and interpretation of the large, complex data sets obtained from a particular sample.

It also requires PCR amplification step to increase the signal to noise ratio; as a result, amplification biases may be introduced, giving rise to spurious CNVs. Moreover, CNV predictions by SNP arrays vary depending on the reference set and computational approach used. [98, 127-138]

SNP-CGH analysis: This technique is designed for genome-wide association studies (GWAS) to form a virtual karyotype. SNP-CGH is a related microarray technology that uses oligonucleotide probes corresponding to allelic variants of selected SNPs. The advantage is that it uses hybridization of genomic DNA to both probe variants to identify heterozygosity, while a signal for only one allele indicates either homozygosity or LOH [139].

Combined MC/SNP array karyotyping: This technique has conclusively shown a higher diagnostic yield of chromosomal defects (74 vs. 44%; p < 0. 0001), compared with cytogenetics alone, often through detection of novel lesions [140].

Gene expression (cDNA) arrays: cDNA microarrays technology measures in parallel many thousands of gene-specific mRNAs in a single tissue sample. The principle of Gene Expression Arrays is comparable to Array CGH (apart from using oligonucleotide). The Affymetrix Gene Chip arrays are most popular cDNA arrays that have been used to classify pathological subgroups of leukemias and potentially prognostic subgroups of melanomas by relative expression levels of a panel of genes.

A special statistical analysis is needed to avoid incidental significances. Permax test was used using a program that is available free on-line. The t-statistics have a tendency to preferentially select genes with very small intragroup variances. After determining the most significant genes from the t-statistics (Permax <0.5), those genes with absolute differences between means ≥100, and ratios of means ≥3 were identified. This selection process provides a conservative determination of genes whose expression levels are changed in different conditions.

The top up regulated and down regulated t-test-ranked genes are identified. Gene expression array technology has been used to study neoplastic transformation in endometrial (pre)cancer. 100 genes were found which are hormonally regulated in normal tissues, are expressed in a disordered and heterogeneous fashion in cancers, with tumors resembling proliferative more than secretory endometrium. [141-147]

Array Painting

Although aCGH and SNP arrays have been used successfully to detect copy number changes in tumors, they are incapable of detecting balanced rearrangements, such as reciprocal translocations or inversions. Array painting has been developed to map such rearrangements at a high resolution on a genome-wide scale. Briefly, in array painting, the two derivative chromosomes from a balanced translocation are separated by flow sorting. They are then PCR amplified, differentially labelled, and hybridized to an array of genomic clones. Only clones corresponding to the sorted chromosomes will show fluorescence above background. Fluorescence ratios of the signals from the two derivatives can be used to identify clones that span a rearrangement breakpoint; as clones have intermediate ratios representing the hybridization of both derivatives [148].

Spectral karyotyping (SKY): SKY gives all chromosomes a unique fluorescent color by hybridizing with chromosome-specific probes each labelled with a different combination of fluorochromes. The advantage with SKY is therefore especially useful for the genome-wide detection of structural chromosomal changes (contrasting CGH, which yields numerical data only). Translocations are readily visible. SKY can screen genome-wide, and does not require prior knowledge of chromosomal breakpoints [25].

Digital karyotyping (DK): This technique is recently been developed for a genome-wide analysis of DNA copy number alterations at high resolution. The principle of this approach is based on the isolation and enumeration of short sequence tags (21 bp each). These tags contain sufficient information that allows assigning the tag sequence to their corresponding genomic loci from which they are derived. After isolation the tags are ligated to each other and are cloned into bacteria. Every bacterial clone represents homogeneous plasmid that contains a certain number of different tags (approximately 32 tags). Generally, about 5000 clones are sequenced from each tumor sample to establish a digital karyotyping library that collects a total of 160,000 tags (32 × 5000). The number of each unique tag along each chromosome can be used to quantitatively evaluate DNA content in tumor samples. Digital karyotyping identified all the known chromosomal alterations including whole chromosome changes, gains or losses of chromosomal arms, and interstitial amplifications or deletions in both cell lines [149-151]. Digital karyotyping (DK) is a high resolution method for genome-wide analysis of copy number changes and other genome rearrangements. In DK, genomic DNA is digested with a mapping restriction enzyme, originally SacI (with a 6 bp recognition sequence) followed by the ligation of biotinylated linkers and a second digestion using a fragmenting restriction enzyme with a 4 bp recognition sequence. The biotinylated sequences are isolated by binding to streptavidin and the DNA tags are released using a tagging enzyme with a 6 bp recognition sequence. The isolated sequence tags are concatenated, cloned, sequenced, and aligned to a reference genome assembly, providing a copy number estimate at the particular locus. The combination of the mapping and fragmenting enzymes used determines the size of detectable rearrangements, and the genome-wide occurrence of mapping enzyme recognition sites defines genomic areas represented in DK analysis. In the case of SacI, recognition sites are abundant and expected to occur every 4 kb; however, some areas of the human genome (<5%) have lower densities of SacI sites and thus would be analyzed at a lower resolution [149]. To date, DK has been successfully applied to the analysis of human ovarian and colorectal cancers, as well as human cancer cell lines from melanoma and medulloblastoma. The method is promising and has been used to detect genome rearrangements, particularly small amplicons of less than 1 Mb and homozygous deletions, previously missed by studies using SKY and CGH. The original version of DK has a theoretical resolution of 4 kb defined by the genomic spacing of SacI sites, which is higher than the available array-based methods. However, DK provides more robust readouts of copy numbers because it depends on digital sequence tag counts rather than hybridization signal intensities produced by array technologies.

A partial limitation of DK imposed by the use of restriction enzymes is the uneven coverage of the genome, which may be addressed by using different combinations of mapping and fragmenting enzymes. Another current limitation is the cost of sequencing, which will be improved with the advent of next-generation sequencing technologies. [152-157]

Immunosorting to isolate tumor cell DNA: Magnetic beads are bound to antibodies that react to the tumor associated antigen on cancer cells. The magnet-isolated tumor cells can be directly used to isolate genomic DNA and RNA or they can be cultured for a short term to further expand the tumor cell population. The epithelial origin of the immunosorted carcinoma cells can be confirmed by staining the purified cells with a cytokeratin antibody or by loss of heterozygosity assay.

Restriction landmark genomic scanning (RLGS): RLGS is a 2D gel electrophoresis method that allows detection of DNA methylation in human tumors if a methylation sensitive landmark enzyme, Not1 is used. In a single RLGS profile, up to 2000 end labelled Not1 sites are displayed. More than 90% RLGS fragments that are cloned are CpG islands. Another way to study DNA methylation is to assay for the level of 5'-methlycytosine content in tumor cells using HPLC analysis [158, 159].

PCR: Polymerase Chain Reaction is a revolutionary method developed by Kary Mullis in the 1980s. PCR is based on using the ability of DNA polymerase to synthesize new strand of DNA complementary to the offered template strand. Because DNA polymerase can add a nucleotide only to a pre-existing 3'-OH group, it needs a primer to which it can add the first nucleotide. This requirement makes it possible to delineate a specific region of template sequence that needs to be amplified. At the end of the PCR reaction, the specific sequence is accumulated in billions of amplicon copies. PCR-based methods have been used to detect known genome rearrangements, particularly alterations in gene copy number. Variations of PCR include, Multiplex PCR, Real-time quantitative PCR nested PCR, Emulsion PCR etc.

Real-Time Quantitative PCR

In real-time quantitative PCR, the accumulation of amplified products is monitored by measuring the fluorescence of probes or intercalating dyes introduced into the reaction. The number of cycles necessary to attain a particular DNA concentration (threshold concentration) is measured as a function of fluorescence intensity. Initial DNA concentrations are determined using the amplification efficiency of each cycle. Real-time PCR has been applied to the detection of specific copy number changes, for example, MYCN amplifications in neuroblastoma. This method is rapid and does not require a large amount of starting material. If whole genome sequencing identifies mutations, RTPCR can be used as a diagnostic tool as it is more amenable to bed side evaluation. It has a very limited throughput, is quite costly, and is unsuitable for the detection of translocations or inversions. In addition, it is unsuitable for genome-wide screens for rearrangements. [160-163, 122]

Multiplex PCR

Multiplex PCR methods, including multiplex ligation dependent probe amplification (MLPA) multiplex amplifiable probe hybridization (MAPH) and more recently, nonfluorescent multiplex PCR coupled to high performance liquid chromatography (NFMP–HPLC), have been designed to screen for copy number changes at multiple loci simultaneously. These methods are more efficient than standard PCR because they allow concurrent screening for rearrangements at multiple loci. However, they are still limited to detecting known unbalanced rearrangements at a few loci at a time. Recently; a single-molecule haplotyping assay has been developed for the genome-wide detection of inversions. This method uses fusion PCR performed on single molecules of genomic DNA. The fusion PCR procedure juxtaposes single-copy sequences on either side of putative inversions in an orientation-specific manner, and inversions are then detected from the haplotypes of these sequences [164-167].

Emulsion PCR: In emulsion PCR, individual DNA fragment-carrying streptavidin beads, obtained through shearing the DNA are amplified. Fragments are attached to the beads using adapters, and are captured into separate emulsion droplets. The droplets act as individual amplification reactors, producing 10^7 clonal copies of a unique DNA template per bead. Each template-containing bead is subsequently transferred into a well of a picotiter plate and the clonally related templates are analyzed using a pyrosequencing reaction. The use of the picotiter plate allows hundreds of thousands of pyrosequencing reactions to be carried out in parallel, massively increasing the sequencing throughput [168, 169].

Sequencing Approaches

DNA sequencing is the process of determining the precise order of nucleotides within a DNA molecule. It includes any method or technology that is used to determine the order of the four bases—Adenine (A), Guanine (G), Cytosine(C), and Thymine (T)—in a strand of DNA. Advent of rapid DNA sequencing methods has greatly accelerated biological and medical research and discovery. Knowledge of DNA sequences has become indispensable for basic biological research, and in numerous applied fields such as diagnostic, biotechnology, forensic biology, and biological systematics.

The rapid speed of sequencing attained with modern DNA sequencing technology has been instrumental in the sequencing of complete DNA sequences, or genomes of numerous types and species of life, including the human genome and other complete DNA sequences of many animals, plants, and microbial species. Escherichia coli alanine tRNA was the first nucleic acid molecule to be sequenced by Holley and co-workers in 1965. Almost all sequencing technologies are used for research purpose and not applied in clinical diagnosis, however the rate at which technologies and discoveries are made, the day may not be too far when specific panels are designed and probed.

Sanger Sequencing

Since its initial report in 1977, the Sanger sequencing method has remained conceptually unchanged. The method is based on the DNA polymerase-dependent synthesis of a complementary DNA strand in the presence of natural 2'-deoxynucleotides (dNTPs) and 2',3'- dideoxynucleotides (ddNTPs) that serve as non reversible synthesis terminators (170). The DNA synthesis reaction is randomly terminated whenever a ddNTPs is added to the growing oligonucleotide chain, resulting in truncated products of varying lengths with an appropriate ddNTPs at their 3' terminus. The products are separated by size using polyacrylamide gel electrophoresis and the terminal ddNTPs are used to reveal the DNA sequence of the template strand. Originally, four different reactions were required per template, each reaction containing a different ddNTPs terminator, ddATP, ddCTP, ddTTP, or ddGTP. However, advances in fluorescence detection have allowed for combining the four terminators into one reaction by having them labelled with fluorescent dyes of different colors [171, 172]. Subsequent advances have replaced the original slab gel electrophoresis with capillary gel electrophoresis, thereby enabling much higher electric fields to be applied to the separation matrix. One effect of this advance was to enhance the rate at which fragments could be separated [173]. The overall throughput of capillary electrophoresis was further increased by the advent of capillary arrays whereby many samples could be analyzed in parallel. These and many other advances in sequencing technology contributed to the relatively low error rate, long read length, and robust characteristics of modern Sanger sequencers.

An inherent limitation of Sanger sequencing is the requirement of in vivo amplification of DNA fragments that are to be sequenced, which is usually achieved by cloning into bacterial hosts. The cloning step prone to host-related biases, is lengthy, and is quite labour intensive [174]. Capillary-based DNA sequencing ('first-generation sequencing', also known as Sanger sequencing): These technologies provided the ability to analyse exonic mutations and copy number alterations and have led to the discovery of many important alterations in the cancer genome [175]. Sanger sequencing, however, was deemed too labour-intensive and costly to allow extensive genome analyses of large numbers of tumors.

Several sequencing platforms evolved from the original Sanger method through innovations in the amplification, sequencing, and detection steps. These second-generation technologies marketed in recent years have vastly expanded the capabilities and applications of DNA, RNA, and chromatin analyses.

Second Generation/Next Generation Sequencing (NGS)

Roche/454: 454 Life Sciences Corp. developed the first NGS technology and fundamentally changed perceptions of what might be achieved with sequencing and in 2010, the first NGS human genome was published using 454 sequencing [168, 176]. Libraries are prepared by ligating oligonucleotide adapters to fragmented genomic DNA. The 454 technology, the first next-generation sequencing technology released to the market, circumvents the cloning requirement by taking advantage of a highly efficient in vitro DNA amplification method known as emulsion PCR [177, 178]. After PCR, the emulsion is broken, and DNA-coated beads are purified, denatured and loaded into the wells of a 'picotiter' plate. The wells of the picotiter plate are large enough for only a single bead to be loaded; each well carrying a bead will generate an individual DNA sequence. Pyrosequencing is performed by cyclical addition of individual nucleotides, sulfurylase and luciferase.

Pyrosequencing: It is a form of sequencing in which at end of synthesis chemiluminescence occurs due to the release of a pyrophosphate group when the correct nucleotide is incorporated [179, 180]. As each nucleotide is incorporated into the growing strand, an inorganic pyrophosphate group is released and converted to ATP by the sulfurylase. Luciferase uses the ATP to convert luciferin to oxyluciferin, producing a light signal that is directly proportional to the number of inorganic pyrophosphate molecules released and the number of nucleotides incorporated. The first publication generated 250,000 reads of 80-120 bp in length [123]. Around 1 million sequences of up to 700 bp in length are currently generated in a GS-FLX (Roche) single run. A recent study, using the 454/Roche sequencing technology, showed the potential of next-generation sequencers to detect rare variants present in specific subpopulations of cells that elude cost-effective detection by capillary sequencing approaches [181].

Illumina method: sequencing by synthesis: The Illumina HiSeq is the most widely adopted NGS instrument to date and was used to sequence the first cancer genomes [182, 183]. Illumina have significantly refined the sequencing-by-synthesis technology it acquired from Solexa in 2006, by improving chemistry, instruments and software. It achieves cloning-free DNA amplification by attaching single-stranded DNA fragments to a solid surface known as a single-molecule array, or flow cell, and conducting solid-phase bridge amplification of single-molecule of DNA templates (Illumina, Inc.) [184-186]. Libraries are prepared by ligating Y-shaped oligonucleotide adapters. These are prepared from two oligonucleotides that share complementarity at one end; when annealed and ligated to DNA fragments. They allow different sequences to be added to the end of each fragment. The library of DNA fragments is enriched by PCR ready for clustering and sequencing. Libraries are denatured to pM concentration and are introduced to an Illumina flow cell; the fragments hybridize to complementary oligonucleotides on the surface of the flow cell and are copied by DNA polymerase. These daughter molecules are then 'bridge-amplified' by repeated cycles of chemical denaturation and polymerase extension to produce discrete clusters each containing about 1000 molecules. SBS uses fluorescently labelled and reversibly blocked terminator deoxynucleoside- triphosphates in a cyclic sequencing reaction. Nucleotides are incorporated by DNA polymerase into the growing DNA strand, the flow cell is imaged to determine which nucleotide has been incorporated into each individual cluster, and finally the terminator is removed by chemical cleavage ready for the next round of incorporation, imaging and cleavage [187]. The early Solexa-based sequencers from Illumina generated reads of 35 bp in 2007 and generated around 30 million sequences or 1 Gb of data from a flow cell. Read length has increased to 150 bp on the HiSeq 2500 system, which generates over 1. 5 billion sequences (or 3 billion paired-end sequences) and 300 Gb of data from a single flow cell as of December 2012. The Illumina approach is more effective at sequencing homopolymeric stretches than pyrosequencing. However the limitation is that it produces shorter sequence reads and hence cannot resolve short sequence repeats. In addition, due to the use of modified DNA polymerases and reversible terminators, substitution errors have been noted in Illumina sequencing data [188].

Life Technologies: Sequencing by Ligation: ABI/SOLiD

Life Technologies initially developed the Agencourt Personal Genomics support oligonucleotide ligation detection (SOLiD™) sequencing technology in their SOLiD 3, 4 and 5500 instruments, but these have not seen widespread adoption by the sequencing community due to reduced throughput and a more complex workflow. The SOLiD system uses emulsion PCR to generate template beads for sequencing by ligation. Beads are then deposited onto a slide and primers hybridize to the adaptor sequence on the template beads. Four fluorescently labelled probes compete for ligation to the sequencing primer. Multiple cycles of ligation, detection and cleavage are performed, with the number of cycles determining the eventual read length of up to 75 bp. However, it uses a sequencing-by-ligation approach based on the sequential annealing and ligation of degenerate oligonucleotide probes, specifically octamers containing known dinucleotide bases. There are a total of 16 distinct octamers (each containing one of the possible 16 dinucleotide sequences); one of 4 fluorescent dyes is used to label 4 of these dinucleotide probes. After annealing to the DNA fragment immobilized on a glass support, the probe is ligated to the preceding nucleotide and the fluorescent label is detected. The octamer is then cleaved between the fifth and sixth bases, thereby removing the label for the next round of annealing and ligation. Using this system, bases are no longer sequentially interrogated but are rather identified at evenly spaced intervals; the complete sequence is generated from an overlapping series of successive ligation events [179, 180, 189, 190]

There are several limitations associated with PCR-based sequencing platforms. They introduce inevitable biases during amplification, as well as errors due to imperfections of the PCR process. Also, optimal PCR conditions differ between sequences with variable base composition, possibly altering the relative amounts of amplified DNA fragments and skewing the frequencies of sequenced bases. Other shortcomings include short read lengths and reduced accuracy in highly uniform homopolymer regions. However, newer versions of these next-generation sequencers will improve on the output speed, read lengths, and sequencing accuracy [179].

Whole genome sequencing (WGS): The first whole cancer genome sequence was reported in 2008, a description of the nucleotide sequence of DNA from an acute myeloid leukemia compared with DNA from normal skin from the same patient [191]. Point mutations, insertions, deletions can be detected. The major potential of whole-genome sequencing for cancer is the discovery of chromosomal rearrangements. In addition to rearrangements between unique, align able sequences, whole-genome sequencing may be able to detect other types of genomic alterations that have not been observable using previous methods. Among the most important of such events are somatic mutations of non-coding regions, including promoters, enhancers, introns and non-coding RNAs (including microRNAs), as well as unannotated regions. The WGS methods could include :a) Deep single end whole genome sequencing-b) Deep paired end whole genome sequencing that is used to detect point mutations, insertions, deletions, amplifications, inter chromosomal rearrangements c) by shallow paired end and deep paired end whole genome sequencing aberrations like insertions, deletions, amplifications, inter chromosomal rearrangements, duplications, inversions are detected. But this method does not preserve genome structure and can give rise to artifactual nucleotide sequence alterations [192]. Thus, whole-genome sequencing provides the most comprehensive characterization of the cancer genome but, as it requires the greatest amount of sequencing, it is the most expensive [193, 194]. Long-distance mate-pair sequencing: is yet another variation of paired end sequencing involving circularization of long DNA fragments (>1 kb) and ligation to an internal adapter DNA of known sequence Leary et al. [195] Shotgun sequencing technique is sufficient to identify somatic rearrangements in the genome and copy number alterations [196, 197].

Other Applications of Next-Generation Sequencing Technology

The high-throughput capabilities of next generation sequencing and its power to conduct genome-wide surveys in an unbiased, high resolution manner have made it a valuable tool in many other applications. In addition to providing reliable maps of SNVs, small insertions, deletions, and copy numbers, it is also instrumental in the search of complex chromosomal rearrangements in solid tumors

Epigenome: Study of the epigenome has made progress with the advent of next generation sequencing, which can replace microarray based assays as a method for the survey of genome-wide DNA methylation. This method uses MethylC-Seq to obtain detailed maps of cytosine methylation at single-base resolution, marked differences in methylation patterns were observed between the genomes of fetal fibroblasts and human embryonic cells. The latter contained a high degree of methylated residues in a non-CG context, suggesting a unique methylation mechanism in pluripotent stem cells [198, 199].

ChIP –Seq: Chromatin immunoprecipitation (ChIP) followed by next-generation sequencing (ChIP-Seq) is now employed to evaluate transcription factor binding sites as well as to characterize variations in histone modification and chromatin structure. This technique successfully demonstrated the epigenomic differences defining distinct subtypes of leukemia [200].

Transcriptome Sequencing: Expression analyses can now be done by next-generation sequencing of SAGE (serial analysis of gene expression) tags or by whole transcriptome sequencing of RNA material (RNA-Seq). In particular, RNA-Seq offers a rapid method of cellular RNA analysis to study the various aspects of the dynamic transcriptome. Applications include detection of splice isoforms, fusion transcripts and miRNA. It also allows one to corroborate findings from genomic studies to verify or identify somatic changes, to annotate novel genes, and to understand the impact of mutations on RNA stability [201-206].

Exome Sequencing/Exome Capture Analysis: A random library of genomic fragments is prepared and platform-specific adaptors are attached to each end. These fragments are then combined with a set of probes that define the human exome, followed by hybridization. The probe: genomic library fragment hybrids are captured using magnetic beads and isolated from solution by application of a magnet, or a solid phase capture. Denaturing conditions are used to elude the captured genomic library fragment population from the hybrids, and prepared for sequencing [207]. The advantage here is the selective sequencing of the coding regions of the genome as a cheaper but still effective alternative to WGS. However Exome sequencing is only able to identify those variants found in the coding region of genes which affect protein function. It is not able to identify the structural and non-coding variants associated with the disease.

Clone-Based Approaches to Rearrangement Detection

Clone-based methods have been recently developed to detect both balanced and unbalanced genome rearrangements. Mapped paired-end sequence reads generated from a human fosmid library to the human reference genome sequence to discover and catalogue structural genomic variants. The basis for this analysis was the tight regulation of insert sizes of fosmid clones imposed by lambda phage packaging machinery; if no rearrangement was present in a clone, the pair of its end sequences would align to the reference sequence with approx 40 kb spacing. Significant deviations in the spacing were indicative of a rearrangement in the clone. The investigators were capable of detecting finer-scale structural variants, such as small indels and inversions, which would not have been detected using conventional array methods. The resolution of this analysis, determined by the insert size of fosmid clones, as well as by the level of read coverage, ranged between 8 and 40 kb [208, 209]

ESP

End sequence profiling (ESP), has been developed and successfully applied to the genome-wide analysis of rearrangements of the MCF7 breast cancer cell line. In ESP, a BAC (Bacterial Artificial Chromosome) library is constructed for the tumor genome of interest, both ends of BAC clones are sequenced, and the paired-end sequences are mapped back to a reference genome assembly. Structural genomic variants are discovered by identifying clones whose paired-end sequences map to the reference genome in orientations that suggest the clone was derived from rearranged DNA. The advantage of ESP approach is that it is potentially applicable to the detection of all types of genome rearrangements that could be inferred from different types of "ESP signatures" [210, 211].

Powerful, paired-end clone sequencing has several limitations.

1. The approach is dependent on the construction of clone libraries, which can be a slow and costly process.
2. The resolution of paired-end sequencing methods is determined by the clone properties and the redundancy of genome coverage. While insert sizes of cosmid and fosmid clones are tightly bounded by the lambda packaging limits of 32–48 kb, BAC insert sizes are less

constrained. This feature of fosmid clones, exploited by Tuzun et al. (2005), may facilitate rearrangement detection; however, a larger number of fosmid clones than BAC clones would be required to achieve a similar degree of genome coverage. Also, with both clone types, large numbers of clones would be necessary to achieve genome-wide high resolution coverage of rearrangements, because the sampling occurs only from the ends.

BAC clone fingerprint profiling (FPP): To address limitation in ESP, recently a BAC clone fingerprint profiling (FPP) approach was developed for the high resolution detection of genome rearrangements, which achieves redundancy of genome coverage with fewer BACs than would be required by end sequencing. The FPP method includes the digestion of genomic BAC clones prepared from tumor DNA with five restriction enzymes, HindIII, EcoRI, BgIII, NcoII, and PvuII to generate clone fingerprints that are then aligned against the in silico digests of the reference genome sequence using the FPP alignment algorithm. Differences between the experimental and in silico digestion patterns are indicative of genomic differences, including genome rearrangements in the clone versus the reference genome. For instance, an alignment in which the clone maps to one genomic region, but in which there are internal gaps in fragment alignments, indicates the presence of a localized rearrangement confined to the clone; on the other hand, an alignment in which the clone fingerprint is partitioned over several regions in the genome suggests the presence of a translocation, inversion, or a large deletion. The FPP approach provides several important advantages over ESP and other genome-wide methods for rearrangement detection.

1.   The method samples the entire clone insert and not just the clone ends, as in ESP. Therefore, rearrangement coordinates mapping within the clone will be more precisely localized with FPP than ESP, given the same number of clones sampled.
2.   FPP is relatively tolerant of repeats compared with ESP and oligonucleotide microarrays, since only 7% of human repeats are found in contiguous regions of 3. 9 kb (the average sizeable HindIII restriction fragment). This is an important advantage, considering that a significant portion of the human genome is composed of repeat sequences [212].
3.   Both balanced and unbalanced rearrangements are potentially detectable.

4. Clones harbouring rearrangements can be directly selected for functional analyses and sequencing (as in ESP).

The main limitation is the cost and speed of library production, the cost of clone characterization, and the requirement of a large amount of starting DNA material. Consequently, although the FPP approach is potentially very powerful, the reliance on clones currently limits its widespread application. FPP may erroneously interpret restriction fragment length polymorphisms as genome rearrangements. This limitation may be partially addressed in the future as more complete catalogues of normal genomic variation are compiled. With the advent of next-generation sequencing technologies, which do not require clones, genome and transcriptome resequencing studies are becoming more and more cost effective, making the prospect of a $1,000 genome a definite possibility in the future [213].

Third-Generation Sequencing

In addition to recent improvements in NGS platforms, new "third-generation" sequencing technologies have also emerged. These technologies detect the binding of nucleoside triphosphates to the polymerase in real time [Pac Bio (Pacific Biosciences), nanopore (Oxford Nanopore Technologies)] and allow the sequencing of nucleic acids from single molecules, thereby circumventing prior DNA amplification and labelling in the library-preparation steps [214].

HELICOS: The HeliScope eliminates the amplification step and directly sequences individual labelled DNA strands bound to a flow cell. Sequencing by synthesis is performed using a reduced processivity of DNA polymerase and labelled nucleotides added to the reaction one at a time. Each incorporated nucleotide is read with a highly sensitive photon detection system called total internal reflection fluorescence, followed by removal of the label and subsequent synthesis cycles. It performs with read lengths of 25–35 bp and an output of 21–35 Gb per run (>1 Gb/h) (http://www. helicosbio.com). However, the HeliScope was not widely adopted and does not have a large installed base in the sequencing arena [179].

Pac Bio RS: A novel third-generation platform commercialized by Pacific Biosciences based on single molecule, real-time (SMRT) technology. In this system, aperture chambers measuring nanometers in diameter called zero mode waveguides (ZMWs) are created in a 100 nm metal film to allow the selective passage of short wavelengths. Each ZMW has a single DNA polymerase attached to the supporting substrate. The fluorescence of each phospholinked nucleotide is detected in real time as it is incorporated into the growing DNA strand, enabling high detection resolution to nucleotide level. The Pac Bio RS utilizes an array of ZMWs that allow simultaneous sequencing of 75,000 DNA molecules in parallel, and is capable of producing long read lengths (average >1,000 bp, up to >10,000 bp) at a fast rate (<45 min per run) (http://www.pacificbiosciences.com) [179].

Ion Torrent: A more recent, approach, is Ion Torrent technology. In Ion Torrent semiconductor sequencing which is the latest technology introduced, detects hydrogen ions that are released as the nucleotides are incorporated into a growing DNA strand [215]. It is based on the detection of pH changes by an ion sensor during DNA synthesis; the output is directly converted and digitally recorded by a semiconductor chip. Ion Torrent technology is the operating basis of the new sequencer, the Personal Genome Machine, which requires as little as one hour per run and will likely decrease the costs of sequencing applications (http://www.iontorrent.com).

Proteomic Biomarkers of Malignant Cells

Proteins have been one of the convenient biomarker used in laboratory medicine due to its ease of detection and also because it is mostly detected in serum, which is not difficult to obtain. However, other sample sources like biopsy tissue both fresh & fixed, body fluids, sputum urine etc., have also been used as clinical as well as research specimens to analyse proteins. Disease pathways are understood through proteins, its changes in structure, function and temporal expression.

The human body expresses about 100,000 to 10,000,000 proteins and only a small percentage of this has been described in terms of its structure and function. The task of elucidating the structure, function, post translational modifications and its interaction with other proteins/DNA/RNA/other small molecules is enormous. The advent of technologies in proteomics such as 2 D gels, automated isolation, extraction of resolved proteins their identification by mass spectrometry have definitely enabled our understanding and function of some proteins associated in cancer.

Other protein identification/characterisation techniques available are Mass spec(MS), HPLC, capillary array(CA), MALDI (matrix associated laser desorption/ionisation), MALDI-TOF (time of flight), TOF-MS, MALDI ion trap, trap-TOF –MS, ESI (electron spray ionisation), Protein array, FRET, Surface enhanced laser desorption/ionisation, tissue microarray etc. [216, 217]. Details of each of these techniques and its impact is not being discussed here, only the proteins identified by these techniques and its importance as a cancer biomarker or as a putative biomarker is being mentioned here as the focus is on genomics rather than proteomics.

Monoclonal Antibodies

Monoclonal antibodies (mAbs) can be used both for diagnosing and targeting cancer. An example of this is a highly sensitive and specific mAb, SF-25, which has been developed against a target protein expressed on the surface of cancer cells, including those of the colon, liver and pancreas. Monoclonal antibody SF-25 represents an excellent potential biomarker of pancreatic adeno carcinoma, and could be configured in an immunoassay for detecting pancreatic adeno carcinoma cells in biological fluids [218]. Commercial products based on SF-25 are in development. Radio labeled mAbs are widely used in the detection and treatment of cancer. There are several approved preparations, such as ibritumomab tiuxetan (Zevalin®; Biogen Idec, Cambridge, MA, US): a mAb conjugated with a radio nucleotide.

Nanobiotechnology-Based Diagnostics and Therapeutics

The ability to control both wavelength-dependent scattering and absorption of nanoshells offers the opportunity to design nanoshells, which provide both diagnostic and therapeutic capabilities in a single nanoparticle. Nanoshells are spherical in shape and consist of a core of non-conducting glass that is covered by a gold shell. A nanoshell-based alloptical platform technology can integrate cancer imaging and therapy applications. Immunotargeted nanoshells, engineered to scatter light in the near infrared range, enabling optical molecular cancer imaging, and to absorb light, can achieve selective destruction of targeted carcinoma cells through photo thermal therapy [219]. In a proof-of-principle experiment, dual imaging/therapy immunotargeted nanoshells were used to detect and destroy breast carcinoma cells that over express HER-2. Use of this combined imaging and therapy approach would facilitate the development of personalised therapy for cancer.

Bioinformatics: The development of faster sequencing platforms necessitates parallel progress in bioinformatic tools to handle, analyze, and interpret the overwhelming amount of sequence data in order to draw meaningful conclusions. Specifically, powerful computational methods are employed in the alignment of multiple sequenced DNA fragments and their reconstruction into an accurate contiguous sequence, which is then analyzed for variants. It is estimated that as much as Sorting Intolerant from Tolerant (SIFT) [220] and Polymorphism Phenotyping (PolyPhen) [221, 222] algorithms are predictive tools used to determine the effects of mutations on protein function. Another useful technique for the discovery of important drivers is an analysis of mutations at the structural protein level (e. g., X-ray crystallography modelling) [223].

The combination of all these approaches will help to filter the large amount of array-based data and allow a focused view of consistent changes. The translational application of whole genome sequencing will make personalized medicine a reality. Each tumor will be sequenced to provide a detailed view of its unique genetic makeup and the forces that have shaped its genome. The ability to conduct accurate multigene profiling and to decipher the genetic signature of tumors, their key defining signalling pathways, and the genetic environment of the individual in which a tumor resides will enable highly effective tailored therapy. All the above mentioned techniques with their advantages and disadvantages have been summarized in Table 1.

Table 1. Comparison of the different technology platforms

Method	Genomic resolution	Advantages	Disadvantages
Conventional			
Chromo-some banding	5-8 Mb (1 Mb = 1 million base pairs)	Detects rare clonal events, genome wide and detects balanced translocation.	Labour intensive, automation is costly, low resolution, only fresh material can be used.
FISH	0. 5 kb	A) WCP: Genome wide screening is possible, intermediate resolution, suitable for automation, detects rare clonal events B) Gene Specific: High	A) WCP: Performed as one-or two-color FISH Translocation Partners may remain unknown B) GS: need to know the gene /locus of interest.

		resolution, Interphase cytogenetics possible, does not depend on proliferating cells	
M-FISH	2-3 Mb	High resolution, Suitable for automation, detects rare clonal events, identifies all balanced and imbalanced aberrations throughout the entire genome.	Cost for mFISH probes, expensive Data processing, resolution depends on the quality of metaphases achievable with fresh material.
CGH	3-10 million base pairs	Provides genome wide screening of chromosomal deletion or amplification with no need for metaphase spreads. It can be applied to both fresh and paraffin embedded material	Balanced structural chromosomal aberrations, such as balanced reciprocal translocations or inversions cannot be detected. Requires sample containing at least 30-35% tumor DNA
aCGH	1 kb-1Mb or better(depends on the size of clones	This technical breakthrough provides a locus-by-locus measure of DNA copy-number changes that significantly overcomes some of the limitations of conventional CGH, high resolution compared to chromosomal CGH	restricted applicability to the detection of genome rearrangements that involve a change in copy numbers
Oligonucle otide aCGH	5–50 kb	Detection- resolution for CNVs, deletions, amplifications, duplications and aneuploidies.	However, small sequence alterations or single base pair mutations will still not be detected, inability to detect aCGH.
ROMA	achieves an average resolution of	The major advantage of using a representation strategy is to minimize	The complexity reduction may lead to unequal representation

	30 kb	the genome complexity and therefore maximize the signal-to-background ratio	of different parts of the genome, potentially leading to erroneous CNV calls.
SNP	~0. 15Mb	Whole genome analysis, ability to detect allelic instability as well as copy number changes, Study of LOH	Challenging to analyze large, complex data sets. Requires costly reagents And specialized equipment
Digital karyo-typing	4kb	DK provides more robust readouts of copy numbers because it depends on digital sequence tag counts rather than hybridization signal intensities produced by array technologies. High resolution.	A partial limitation of DK is imposed by the use of restriction enzymes is the uneven coverage of the genome, which may be addressed by using different combinations of mapping and fragmenting enzymes, Cost of sequencing.
Next Generation Sequencing			
Sangers sequencing	>25 bp	Originally, four different reactions were required per template, each reaction containing a different ddNTP terminator, ddATP, ddCTP, ddTTP, or ddGTP. However, advances in fluorescence detection have allowed for combining the four terminators into one reaction by having them labelled with fluorescent dyes of different colors.	Requires in vivo amplification of DNA fragments that are to be sequenced, which is usually achieved by cloning into bacterial hosts. The cloning step is prone to host-related biases, is lengthy, and is quite labour intensive.
ROCHE 454	Sequencing throughput is 0. 4-0. 6Gb per run, 80-120bp	Detects rare variants present in specific subpopulations of cells.	Chemiluminescent signal intensity is proportional to the amount of

	read length		pyrophosphate released and hence the number of bases incorporated, the pyrosequencing approach is prone to errors that result from incorrectly estimating the length of homopolymeric sequence stretches (i. e., indels)
Illumina	54-60Gb per run, 150bp read length	The Illumina approach is more effective at sequencing homopolymeric stretches than pyrosequencing.	It produces shorter sequence reads and hence cannot resolve short sequence repeats.
ABI/SOLiD	25Gb per run, 75 bp read length	Enables the distinction between a sequencing error and a sequencing polymorphism.	
Third Generation Sequencing			
HELICOS	25-35 bp read length and output of 21–35 Gb per run.	eliminates the amplification step	Does not have a large installed base in the sequencing arena.
PacBio RS	>1,000 bp, up to >10,000 bp read length.	enabling high detection resolution to nucleotide level	A disadvantage of the Pac Bio RS platform is the error rate of 10-18% which is high compared to other sequencing technologies.
Ion Torrent	400 bp per run	requires as little as one hour per run and will likely decrease the costs of sequencing applications	

Pathogenesis and Cancer Biomarkers

Pathogenesis

Pathogenesis could be defined as the origin or the mechanism by which a disease is caused. In the course of its development it may result in acute, chronic or recurrent state of the disease. Often a potential aetiology or pathological link needs to be established as a causative for a disease. In the case of infectious disease the epidemiological data enables this, however in the case of metabolic disorders or malignancies the critical cellular events leading to diseased state needs to be identified. The primary question addressed is why and what is the deviation from normal. The accurate cause needs to be established and correction of this cause is what will ameliorate the condition. A holistic understanding of how cancer is caused is lacking even today, and that is the reason it is still not completely curable and most medical management is palliative. The search for better biomarkers and treatment is an ongoing effort. It is hence necessary to enquire as to, where, when, how and what goes wrong or shifts a normal cellular event to an abnormal cellular event. Control of cell proliferation as well as its rate has been a central enigma in biology since the very beginning. Throughout embryological development as well as through all adult life many differentiated cells have to make the choice of: to divide or not to divide. Understanding of this would enable us to come to grips with the essence of various malignancies, hereditary changes in cells that lead to uncontrolled cell proliferation. During the establishment of cell cultures it had become eminent that cell transformation leading to the establishment of cell lines was accompanied by chromosomal number changes till it finally was stably fixed. Thus, that chromosomal changes are associated with transformation is a very early finding. Today key facts are at last emerging that are providing information on critical processes that control cell proliferation. Early pointers in understanding cell proliferation came from efforts to establish cell cultures in vitro specially peripheral blood lymphocytes and fibroblast cells that established the need of Platelet derived growth factor (PDGF), Insulin like growth factor (IGF), epidermal growth factor (EGF) in bringing about cell proliferation. Others added to the list are TGF (transforming growth factor ALFA, TGF BETA), NGF, (nerve. growth hormone) somatotropin, GHRF (growth hormone releasing factor) [224-234]. At least three types of mechanisms are considered to explain how growth factor –receptor interactions exert pleiotropic effects on cell proliferation These include the generation of ion fluxes, the activation of protein kinases and the enzymatic generation of small hydrophilic molecules that can freely diffuse away from the plasma membrane. These mechanisms operating as growth factors, second signals or messengers (called as Second messenger) when perturbed in the metabolic pathway are known to have dramatic effects on cell growth [235-241].

Table 2. Oncogene Families

Oncogene	Sub-cellular location of protein	Properties/ function of protein
Class I	Protein kinases	
Src	Plasma membrane	Tyrosine specific protein kinase
Yes	Plasma membrane	Tyrosine specific protein kinase
Fgr	Plasma membrane	Tyrosine specific protein kinase
Abl	Plasma membrane	Tyrosine specific protein kinase
fps(fes)	Cytoplasm	Tyrosine specific protein kinase
erbB	Plasma membrane (transmembrane)	EGF Receptor/tyrosine specific Protein kinase
Fms	Plasma membrane (transmembrane)	CSF-1 receptor/tyrosine specific protein kinase
Ros	Plasma membrane (transmembrane)	tyrosine specific protein kinase
Kit	Plasma membrane	
Mos	Cytoplasm	Serine/ threonine protein kinase
raf(mil)	?	Serine/ threonine protein kinase
Class II	GTP binding proteins	
H-ras	Plasma membrane	Guanine nucleotide binding protein with GTPase activity
K-ras	Plasma membrane	Guanine nucleotide binding protein with GTPase activity
Class III	Growth factors	
Sis	Secreted	Derived from gene encoding PDGF
Class IV	Nucleoproteins	
Myc	Nucleus	
Myb	Nucleus	
Fos	Nucleus	
Ski	Nucleus	
Class V	Hormone receptor	
erbA	Cytoplasm	Thyroid hormone receptor
Unclassified		
Rel	?	
Ets	?	

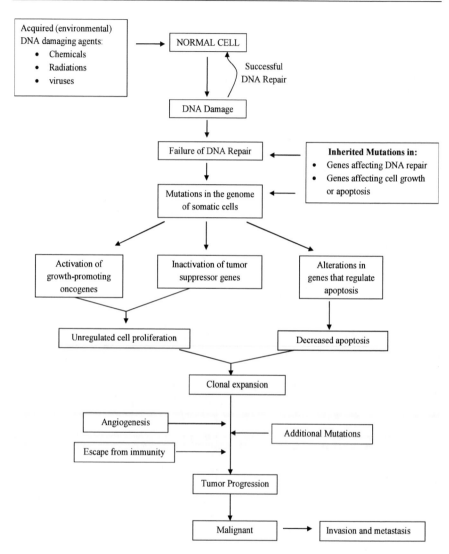

Figure 2. Transformation of a cell from normal to malignant.

The studies of retro viruses that transform cells in culture have enabled the identification of 30 cellular oncogenes, and they could be divided into the following classes. Oncogenes in the same family encode proteins with similar enzymatic activity or intracellular location suggesting their role in transformation may be similar (Table 2).

Cancer therefore is a consequence of multiple changes, occurring both in parallel and in sequence in the macromolecules of the cell viz are DNA, RNA & protein. Earlier only gross end points were possible to be identified and measured, but today with technological advances it has become possible to interrogate events that are very early and at scales hitherto irresolvable. Thus changes in these cellular events are studied at different levels such as: chromosomal, DNA, RNA protein, cellular metabolic and so on. DNA changes in cancer are expressed in terms of gross chromosomal anomalies or could be in the sequence. Major steps involved in the transition of a cell from normal to malignant is depicted schematically in figure 2. A brief account of various mechanisms leading to chromosomal anomalies and hence instability of the cells culminating in a malignant cell is discussed.

Chromosomal Instability in Cancer Cells

Abnormal   chromosomal   content,   characterized   as   changes   in chromosomal structure and number is typical of cancer cells and are generally more numerous in malignant tumors than in benign ones. The karyotypic complexity and cellular heterogeneity observed is often associated with poor prognosis. The functional significance of these and the method of its acquisition by cancer cells are still not clear. The progression to cancer is often associated with instability and the acquisition of genomic heterogeneity, generating both clonal and non-clonal populations. Genetic, or genomic, instability refers to a series of genetic changes occurring at an accelerated rate in cell populations derived from the same ancestral precursor. Its plausible role(s) in tumorogenesis has gained attention in the last decade [242-244]. The estimated rate of somatic cell mutation is in the order of 2.0×10^{-7} mutations/gene/cell division [245]. Genomic instability is a general term used to describe the overall processes that increase this rate of mutation, thus enabling cell(s) to develop new and aggressive phenotypes, to adapt to the changing selection pressures. Genomic instability is generally classified into two major types: microsatellite instability (MIN), and chromosomal instability (CIN) [246]. MIN involves simple DNA base changes that occur due to defects in the DNA repair processes including base excision repair, mismatch repair and nucleotide excision repair [247, 248]. Mitotic segregation errors strongly influences copy number, while structural aberrations can occur at unstable, also called as fragile genomic regions, or through aberrant DNA repair or methylation. Combined molecular cytogenetic analyses can evaluate cell-to-cell variation, and define the complexity of numerical and structural alterations. Because structural change may occur independently of numerical alteration, the term structural chromosomal instability [(S)-CIN] distinguishes it from numerical for which CIN is proposed. MIN and CIN mechanisms are generally found to be mutually exclusive and to produce different phenotypes [249, 250], although recent findings suggest there may be some overlap in these two pathways [251].

FISH-based methodologies are ideally suited to study CIN in comprehensively determining the extent and biological contributions of both clonal and non-clonal chromosomal aberrations to cancer cell populations.

In addition to addressing numerical change, CIN has also been used to describe the presence of a high-level of aneuploidy in a tumor with a complex karyotype. Thus, the "CIN phenotype" may more loosely describe changes in chromosome copy number, ploidy and the presence of structural alterations. Defective mitotic processes such as anaphase failure, centrosome duplication or aberrant DNA repair may lead to tumors with karyotypes bearing both numerical and structural alterations characterized by complex additions, deletions, and translocations that may be balanced or unbalanced (i. e. bearing numerical copy number change). In addition, there may be specialized classes of chromosomal alterations such as, double minute chromosomes (DMs), homogeneously staining regions (HSRs), ring chromosomes (r), multicentric chromosomes and multi-radial chromosomes that are indicative of CIN. The acquisition and systematic evaluation of structural complexity is a less studied aspect of CIN.

Tumor heterogeneity arising as a consequence of CIN often confounds analyses using conventional bulk extracted nucleic acids. Laser capture micro dissection of tumors is one approach for analyses and interpretation of findings from mixed cell populations [252]. Unfortunately however in this process the means by which the genomic imbalances are represented within the karyotype is lost. This underscores the need for parallel metaphase and interphase molecular cytogenetic studies to complement such high-throughput methods to provide a more precise correlation between genomic alterations and the extent of CIN in a patient sample. In addition to classical cytogenetic analysis such as Giemsa banding (G banding), the use of diverse FISH-based methods such as spectral karyotyping (SKY) [253], multicolor FISH (MFISH) [254] and MBanding (MBand) [255] can now address different aspects of CIN to comprehensively describe the cell-to-cell variation, and the genomic complexity of tumor karyotypes in considerable detail. Regions with a high GC content are associated with translocations while AT rich regions predominate deletions, these regions are also called as hot spots where there is greater association with chromosomal rearrangements and correlation with the disease condition [256].

Alu-Repeat Elements

The Alu motif approximately 300 bp in length and recognised by ALU endonuclease is a well characterized repeat element in the human genome. It comprises about 10% of the human genome that frequently undergoes recombination [257-259]. Studies of the t (9:22) suggest they may have a role in facilitating interstitial deletions at the time of 9:22 chromosome translocation in chronic myeloid leukemia (CML). Based on the findings in the GRaBD database [256], Alu mediated deletions and translocation may occur both through homologous and non-homologous recombination events. Studies into the t (11: 22) translocation [260] have identified a region on 22q11 within a low-copy repeat (LCR22) and within an AT-rich repeat on 11q23, which has been implicated in mediating different rearrangements on 22q11. Similarly, within part of chromosome 14 a novel structural copy number polymorphism (CNP) that appears to have an increased level of low-copy number DNA repeat in some types of tumors [261] has recently been identified.

Repetitive Sequences in Centromeric Regions

It is well established in the cancer cytogenetics literature that chromosomal rearrangements frequently occur in centromeric regions [262, 263]. The spatial position of centromeres in interphase nuclei may be one contributing factor in increasing their propensity to rearrange. There is a strong molecular support for the idea that translocation frequencies are elevated in these regions because of the extensive short tracts of sequence homology that may be shared by the translocated chromosome partners. For example, centromeric regions of chromosomes 1, 9, 16 and Y contain satellite III DNA consensus sequences consisting largely of (GGAAT) n that are interspersed amongst the alpha-satellite DNA [264]. Such high levels of sequence homology may act as recipient and donor sites for translocation [265] and facilitate CIN phenomena called "jumping translocations". These aberrations have been defined as non-reciprocal translocations involving a donor chromosome arm or chromosome segment fused to several different recipient chromosomes [266]. A high frequency of jumping translocations of 1q is seen in multiple myeloma samples. The decondensed pericentromeric heterochromatin identified in these karyotypes suggests that the hypomethylated state of the DNA in these regions, following duplication, permitted the recombination of similar centromeric repeat sequences from other chromosomes. The resulting transient tri-radial formations would then be resolved, resulting in the centromeric translocation. Structural rearrangements in osteosarcoma tumors and cell lines showed that almost 30% of the breakpoints were in the pericentromeric region [267]. In similar studies of prostate cancer [268, 269] jumping translocations were also implicated with a high frequency of breakpoints at centromeric and pericentromeric regions.

The Cancer Fragilome

A structural chromosome aberration are frequently caused by DNA damage at certain regions within the genome, known as common fragile sites (cFS) and is also called fragilome. These cFS are especially prone to forming site-specific gaps or breaks when cells are grown under conditions of replication stress. Instability at cFS is believed to be a critical event in the generation of DNA damage during tumor development. Fragile sites (FSs), span approximately 50 kb to 1 MB, which are prone to breakage /reunion and are sites for the exchange of genetic material between sister chromatids, chromosomal translocation, deletions, gene amplification and sites for the integration of oncogenic viruses [270, 271]. In a metaphase spread, they appear as gaps or breaks in the chromosome, particularly when cells are exposed to ionizing radiation or to chemical compounds inducing physiological stress. There are at least 104 FSs, comprising of about 80 common sites and the remaining sites are more rare [271, 272]. Fragile sites typically cluster at G-light (Giemsa staining-light) regions, which are GC-rich Alu-repeat regions. These regions are also generally associated with a more relaxed conformation of the DNA as well as gene-rich regions. In addition, FSs also centre around CpG islands implicating the role for altered transcription when the fragile site is disrupted. Finally, FSs are also late-replicating, providing the opportunity for breakage and recombination. The role for fragile sites in oncogenesis was first identified in the hematological malignancies [270], where the identification of chromosomal changes were coincident with the location of cancer-specific chromosomal changes and the location of cellular oncogenes . Of the fragile sites identified, several have been widely studied including FRA8C, FRA3B, FRA16D, FRA7G and FRAXB. FRA8C which have been associated primarily with the leukemia's, lymphomas and sarcomas, with its mapping position at 8q24 in the region of the myC oncogene [273]. FRA3B is located at 3p14. 2, which is the most commonly expressed FS among all cancers. FRA3B maps to the region that contains the fragile histidine triad gene (FHIT), shown to be aberrantly expressed in various cancers including lung and cervical cancer (274) and hepatocelluar carcinomas [275]. FRA16D, like FRA8C, is associated with specific and recurrent translocations in leukemias and lymphomas and co-localizes to the WWOX gene. FRA7I and FRA7G have been recently identified to co-localize to the MET oncogene at 7q [276]. Larissa Savelyeva and co-workers have initiated a project (FRAGILOME), with the aim of developing new targets for early cancer diagnosis and individualized therapy. For the detection of genetic aberrations, a dedicated DNA chip was designed; targeting sequences of 21 newly identified cFS-associated genes. Subsequent analysis of different tumor types revealed that a number of these genes had indeed undergone genomic damage. . During recent decades, DNA damage has emerged as a major culprit in cancer. Alternatively, DNA damage may trigger cell death or senescence, contributing to aging [277]. Jan Hoeijmakers and co-workers focused on one potent DNA repair pathway involving nucleotide excision repair (NER), which eliminates helix-distorting lesions. Two NER modes exist: (i) global genome NER (GGNER), which operates genome wide and prevents mainly mutations (and thereby cancer), and (ii) transcription-coupled repair (TCR), which removes damage that obstructs transcription, counteracting cytotoxic effects of DNA injury (thereby delaying aging, but enhancing the risk of cancer). Mutations in the NER helicases XPB and XPD give rise to such syndromes. Different single and double mouse NER mutants exhibit progressive osteoporosis, neurodegeneration, early infertility, growth deficit, liver and kidney aging, deafness, retinal photoreceptor loss, depletion of hematopoietic stem cells, and reduced life spans ranging from 1. 5 years to only 3-5 weeks for double mutants. A striking correlation has been found between the severity of repair defects and the severity of aging manifestations, in support of the DNA damage theory of aging. Conditional mutants revealed organ-specific accelerated aging, including targeted neurodegeneration in specific parts of the brain. Jan Hoeijmakers and co-workers propose that endogenous lesions hamper transcription and replication, triggering apoptosis and premature aging. Persisting DNA damage elicits systemic suppression of the somatotrophic axis and up regulation of defences, favouring maintenance at the expense of growth similar to dietary restriction, which promotes longevity. Together, these data link DNA damage to life span and open up prospects for promoting healthy aging and reducing cancer risk.

Sub-Chromosomal Compartments of the Interphase Nucleus of Cancer Cells

Genes near/at euchromatin permit gene expression but near heterochromatin enable gene silencing. Chromosomes, centromeres and telomeres all have specific spatial orientation within the nucleus. It has been observed that telomeres in tumor cells, under the control of MYCC, formed disorganized aggregates within the nucleus, in contrast to a normal non-overlapped formation. The formation of nuclear aggregates is considered to increase the likelihood of CIN events. Moreover, chromosomal aberrations, both clonal and non-clonal also take place under these circumstances [278, 279].

Contribution of the Breakage Fusion Bridge Cycle (BFB) and Telomere Dysfunction to Chromosomal Structural Heterogeneity

A mechanism called breakage fusion bridge (BFB), is found in highly malignant neoplasm, such as sarcomas of the bone, [267, 280, 281] soft tissue [282], as well as carcinomas of the pancreas [283] and ovary [284]. BFB is a cycle involving chromatid breaks and fusions triggered by dicentric and ring chromosome rupture during anaphase resolution. This self-perpetuating process gives rise to amplifications. Homogenously staining region-HSRs, ladder amplifications, complex chromosomal rearrangements, inverted repeats, interstitial deletions and large duplications. The result of this dynamic process is karyotypic heterogeneity as shown in leiomyosarcomas, malignant fibrous histocytoma, pancreatic carcinomas, ovarian carcinomas, oral squamous cell carcinomas and osteosarcoma [283, 285]. Studies in osteosarcoma cell lines and primary tumors, used the integrated data from SKY [267], M band and aCGH to show that the structural complexity of several clonal and non-clonal chromosomal aberrations were consistent with BFB cycles. Additional instability related studies have observed an association between BFB events, structural chromosomal complexity and telomere length. Telomeres are complex nucleoprotein structures located at the ends of linear chromosomes, critical for maintaining genome integrity [286]. In the absence of telomerase, telomeres progressively shorten. The loss of telomere capping function resulting in dysfunctional telomeres which is one of the telomere-mediated mechanisms promoting genomic instability. Excessive telomere shortening has been shown to lead to BFB cycles and, eventually to generalized genome instability and leading to either cell death or crisis. Only cells that have acquired a telomere maintenance mechanism can escape from this crisis. Telomere length analysis studies [269, 283, 28, 289] have revealed a relationship between telomere length and karyotypic complexity, such that cells with shorter telomeres possessed more structurally complex chromosomal aberrations, including the presence of inverted repeat structures indicative of BFB events. This suggests that despite telomerase expression, the reduced telomere length may be the driving force for the observed BFB events and elevated levels of CIN in prostate cancer, and perhaps other neoplasm [269].

Cellular Events Promoting Genomic Instability

DNA Methylation

Gene expression patterns in cancer cells are altered as a consequence of either mutations in the genes or by epigenetic modifications of the genes or chromosomes. Early studies showed that differential DNA methylation patterns exist in normal and cancer cells [290]. In the human genome, epigenetic modification to DNA by methylation is primarily found in repetitive DNA elements where it is thought to play a role in protecting against spurious recombination events and to silence potentially destabilizing transposable elements. Hypomethylation of the genome results in an increased mutation rate, including genome deletions and chromosomal copy number changes [291, 292]. Direct DNA methylation could occur in the genes or indirectly by methylation, acetylation, or phosphorylation of histones and other proteins around which DNA is wound to form chromatin. DNA methylation at the cytosine residue is the main epigenetic modification in humans which occurs in the context of 5'-CpG-3' dinucleotides. Cancer initiation and progression have been reported as a consequence of epigenetic events. Activity of DNA methyltransferases, (DNMTs) an enzyme that adds methyl groups to cytosine residues are known to be altered in tumor cells. Genomic hypomethylation may lead to both genomic instability and stronger gene expression . Functional silencing of tumor suppressor genes, occurs when local promoter CpG hypermethyaltion is induced. Cell-cycle control and apoptosis genes viz. p14, p15, p16, Rb, DAPK; DNA repair genes MGMT, hMLH1;adhesion and metastasis genes CDH1, CDH13; biotransformation genes GSTP1 and signal transduction genes RARβ, APC which are all genes involved in tumorogenesis are targets for gene silencing. [293-297]

DNA methylation occurs within complex chromatin networks and is affected by histone modifications, which in turn are often disrupted in cancer cells. It has been proposed that distinct histone modifications may form so-called histone code [298]. Accordingly, acetylation of histone lysine's is associated with transcriptional activation, as is methylation of histone H3 at lysine 4 (K4), whereas methylation of H3 at K9 or K27 and methylation of H4 at K20 are associated with transcriptional repression. Histone modifications are reversible processes that are mediated by antagonizing enzymes thus putative drug –drug target. It turns out that each tumor type has its own defined hypermethylome [299]. Furthermore, the expression of histone-modifying enzymes can distinguish cancer tissues from their normal counterparts, and these enzymes may differ according to tumor type [300]. In conformity with this concept, the translocation, mutation, over expression, or amplification of histone modifying genes (e. g., CREB, CBP-MOZ, MLL, NSD1, NSD3, EZH2, MLL, and KDM4C (previously JMJD2C) have been reported in some cancers and also in myelodysplastic syndromes. [299, 301]. In addition, it was recently found that heritable gene methylation may predispose to cancer [302, 303]. Methylation studies in various types of tumors have found it to be tightly connected to cancer development through local hypermethylation of tumor-suppressor genes causing transcriptional repression and/or global genomic hypomethylation resulting in the expression of oncogenes . A causal link was proposed between DNA hypomethylation and aneuploidy in human colorectal cancer cell lines and more recently in primary human colon tumors. CIN is evident in hypomethylated centromeric regions of peripheral blood lymphocyte metaphase cells derived from patients with immunodeficiency and facial anomalies syndrome (ICF) [304-307].

Aberrant DNA Repair Pathways

Nucleotide excision repair, base excision repair and mismatch repair are associated with MIN, whereas elevated structural and numerical aberrations are rarely evident within this group. This would be expected since the MIN phenotype utilizes an intact complementary DNA template when repairing a single strand break. In contrast, DNA repair for the CIN group of tumors typically utilizes homologous recombination (HR) and/or non-homologous end joining (NHEJ) pathways when resolving a double-stranded break (DSBs). DSBs may be caused by several factors including radiation, free radicals arising in hypoxia, DNA damaging agents, telomere erosion and replication errors following a single-stranded break as well as mutations or epigenetic changes of DSB-associated genes. When a DSB is detected, the cell cycle is arrested to allow for the recruitment of repair factors [308]. DSB repair in the S- and G2-phases of the cell cycle is dominated by HR, where replicated DNA can provide an identical copy of the template, so that sequence fidelity is maintained whilst repair is completed. In contrast, the more error-prone NHEJ mechanism dominates the G1-phase and since there is no homologous template available, the integrity of DNA sequence at the site of breakage may be disrupted. In both HR and NHEJ, chromosomal deletions or translocations may result when broken ends from different chromosomes are indiscriminately fused together by error-prone repair mechanisms. Studies using murine models show that defects of the various proteins involved in both pathways may increase the frequency of neoplasia and can elevate the extent of CIN [309]. Components of the HR pathway are associated with BRCA1 and BRCA2 proteins [310]. BRCA1 is known to play essential roles in HR as well as NHEJ, by regulating the expression of genes involved in the repair process [311]. The loss of function of the BRCA1 protein is thought to compromise HR function resulting in the accumulation of chromosome damage and cell cycle abnormalities. Moreover sequence-fidelity recognition defects in a compromised HR pathway may increase the propensity to undergo promiscuous healing between non-syntenic regions and lead to translocations jumping to centromeres or telomeres at high frequency.

Mitotic Segregation Errors

Successful cell division is dependent on the faithful replication of DNA, attachment to the spindle apparatus, alignment along the metaphase plate, segregation and migration to the appropriate poles and cytokinesis. The failure of any of the steps leading into and out of mitosis can cause improper cell division, resulting in the normal execution of apoptotic pathways. The malignant transformation of cells confers the ability to escape normal apoptotic pathways, permitting the survival of abnormal daughter cell(s). Many of the hematological malignancies, which are generally near diploid , the chromosomal complement of many epithelial carcinomas, bone and soft tissue sarcomas ranges between tetraploid (4n+/−) and triploid (3n+/−) implicating mitotic segregation errors as the driving mechanism for numerical instability and aneuploidy. The primary forces that drive a cell to aneuploidy include centrosome duplication, chromosome cohesion defects, merotelic attachments defects of chromosomes and cell cycle/mitotic check point defects. Centrosome duplication has been observed in many cancers including osteosarcomas, ovarian, breast and prostate cancers and have been shown to result in the unequal segregation of genomic material in a multi-polar fashion [312-314]. Merotelic attachment refers to the improper attachment of chromosomes to the spindle microtubules, where one kinetochore is simultaneously attached to microtubules from both poles [314]. Changes in ploidy may be accomplished through the failure of cytokinesis yielding a doubling of the DNA content (as well as the centrosomes), or failure of the G1/S checkpoint.

Metabolic Changes

A prominent and fundamental change in many tumors irrespective of their histological origin and the nature of mutations is, enhanced glucose utilization called the Warburg effect. Mechanisms underlying this fundamental alterations in metabolism during carcinogenesis include mutations in the mitochondrial DNA resulting in functional impairment, oncogenic transformation linked up with regulation of glycolysis, enhanced expression of metabolic enzymes and adaptation to the hypoxic tumor micro-milieu in case of solid tumors. Classification and prognosis of cancers, besides predicting the response to therapy is based on these observations to calculate a bio energetic index of the cell (BEC index). Metabolic status is linked to alterations in cell signalling related to defence against oxidative stress, redox signalling and damage response pathways, particularly the down regulation of mitochondrial dependent apoptosis in tumor cells with enhanced glucose usage and Hexokinase II levels [315-319].

Mammalian target for rapamycin-mTOR is an evolutionarily conserved serine-threonine protein kinase that belongs to the PIKK [phosphoinositide 3-kinase (PI3K)-related kinase] family, and plays an important role in regulating cell growth and proliferation. Upon activation, mTOR increases the phosphorylation levels of its downstream targets that include p70S6K and 4EBP1, which leads to increased levels of translation, ribosome biogenesis, and reorganization of the actin cytoskeleton and inhibition of autophagy. As a result, mTOR activation promotes cell growth and proliferation, whereas mTOR inhibition stops cell growth and initiates catabolic processes, including autophagy. The phosphatidylinositol-3-OH kinase (PI (3) K)–PTEN–mTOR signalling pathway is aberrantly activated in many tumors, leading to dysregulation of cell growth and proliferation [320, 321]. Activation of this pathway can be assessed by biomarkers such as loss of PTEN mRNA or protein production in tumor tissue.

Telomerase

Telomeres are tracts of repetitive DNA (TTAGGG/ AATCCC for human telomeres) that protect chromosomes from degradation and loss of essential genes. Under normal circumstances, telomeres progressively shorten in most human cells with each cycle of cell division and the length in adult human tissues is approximately half that of the new born. Telomerase belongs to a class of enzymes known as reverse transcriptases that use RNA as a template for creating DNA and it contains both RNA and protein components. The enzyme ensures the maintenance of telomere and thereby protecting the cell from degradation and death [322]. Since telomerase is found in nearly 90 per cent of human cancers and is responsible for indefinite growth of cancer cells, it has been a target for anticancer therapeutics that turn-off telomerase and thereby inhibits tumor growth [323, 324]. The levels of telomerase are also elevated in stem cells allowing unlimited division necessary for the repair of damaged and worn out tissues. Most human tumors not only express telomerase but interestingly also have very short telomeres. Telomerase is one of the best markers for human cancer, associated with only malignant tumors and not the benign lesions making it a diagnostic marker as well as an ideal target for chemotherapy [325-327]. In normal cells, telomerase is sequestered in an area of the cell nucleus called the nucleolus, away from the chromosomes. The enzyme is released only when needed during cell division, and then returns quickly to the nucleolus thereafter. In cancer cells, however, telomerase is found throughout the cell, implying that the telomerase-shuttling system is impaired. Identification and manipulation of proteins normally involved in telomerase transfer could prove to be useful targets for anti-telomerase therapies [328].

p53

The p53 gene is one of the tumor suppressor genes that normally prevent uncontrolled multiplication of abnormal cells and experimental findings from the last two decades have established a crucial role for wild-type p53 in intrinsic tumor suppression [329, 330]. Upon stimulation (e. g., by moderate levels of DNA damage), p53 activates molecular processes that delays the cell cycle progression of proliferating cells and simultaneously stimulating DNA repair processes [331, 332]. On the other hand, higher level of damage has been found to activate p53 mediated cell death pathway (typically apoptosis), a mechanism that is purported to be responsible for the prevention of carcinogenesis. During malignant transformation, p53 or p53-pathway related molecules are disabled most often and a mutant form of p53 may not only negate the wild type p53 function but also play an additional role in tumor progression. Nearly 50 per cent of all human tumors carry a mutated p53 gene [333].

Tyrosine Kinase

Tyrosine kinases are a class of enzymes that regulate multiple cellular processes by acting primarily as important transducers of extracellular signals influencing diverse functions such as cell growth, differentiation, migration, and apoptosis that contribute to tumor development and progression. Many human tumors display aberrant activation of tyrosine kinases caused by genetic alterations that could be related to the malignant transformation [334]. The erbB or HER family of transmembrane tyrosine kinase receptors, especially receptors erbB1 (or EGFR) and erbB2 (or Her2/neu), have been identified as important therapeutic target in a number of cancers. Her2/neu, is over expressed in nearly 30 per cent of patients with aggressive breast cancer, while EGFR is over expressed in several solid tumors [335].

Histone Deacetylases (HDACs)

Acetylation of proteins orchestra the dynamic interplay between various processes like repair of DNA damage, cell cycle arrest and apoptosis determining the cellular response to radiation and various chemotherapeutic drugs. This acetylation is catalyzed by histone acetylases (HATs) that uses acetyl-CoA as substrate and the acetyl group is transferred to the ϵ amino group of certain lysine side chains within histones N-terminal tails and other nuclear receptor proteins thereby regulating chromatin remodelling and gene expression [336]. Chromatin remodelling during the regulation of gene expression is orchestrated by a concerted action of HATs and HDACs that condenses and de-condenses the chromatin structure by acetylating and de-acetylating histones and other nuclear receptor proteins. Further, HDACs appear to be closely associated with oncogenesis by regulating the expression of certain tumor suppressor genes leading to excessive proliferation and tumorogenesis [337]. It appears therefore, that HDAC inhibitors with pleiotropic actions in modulating multiple genes, signalling pathways and biological features of malignancy are useful in the treatment of cancers with multiple oncogenic abnormalities targeting the protein acetylation involved in the regulation of cell signalling (338).

Peptidyl-prolyl cis –Trans isomerase NIMA-interacting 1(PIN1)Pin 1, or peptidyl-prolyl cis/trans isomerase (PPIase), isomerizes only phosphor Serine/Threonine-Proline motifs. The enzyme binds to a subset of proteins and thus plays a role as a post phosphorylation control in regulating protein function. Studies have shown that the deregulation of Pin1 may play a pivotal role in various diseases. Notably, the up regulation of Pin1 may be implicated in certain cancers. It is well known that the functional status of many proteins is regulated by kinase mediated phosphorylation and other post-translational modifications. Recently, regulation of proteins beyond phosphorylation has been unravelled , which is ,in the form Cis and Trans isomerisation (a post-phosphorylation event) of phosphor-serine/threonine - proline peptide bonds, at selective sites catalyzed by peptidyl-prolyl isomerase (PPIase), Pin1 [339, 340]. These conformational changes can have profound effect on the function of proteins, modulating their activity, phosphorylation status, protein-protein interactions, sub-cellular localization and stability. Over expression of Pin1 has been reported in human breast cancer cell lines and tissues, and its expression closely correlates with the level of cyclin D1 (important cyclin required for cell proliferation) in tumors [341]. Pin1 over expression not only confers transforming properties on normal mammary epithelial cells, but also enhances transformed phenotypes of Neu/Ras-transformed mammary epithelial cells and is implicated in mitotic regulation. In contrast, inhibition of Pin1 suppresses the Neu- and Ras induced transformed phenotypes or induces tumor cells into mitotic arrest and apoptosis [342, 343]. Inhibition of Pin 1 through various approaches, such as mutations, deletions or expression of antisense, induces mitotic arrest and apoptosis in tumor cell lines [344]. It appears that Pin1 can be used as a diagnostic marker for the detection of the cancer or to stage the disease, albeit in only certain types of cancers. Recently potential Pin1 sites on topoisomerase IIα (a vital nuclear enzyme and mitotic protein) have been identified, and shown that the two proteins functionally interact with each other resulting in the activation of Topo IIα [345, 346]. Further understanding on the role of Pin1 in tumorogenesis is required before its use as a target for developing antagonists ensuring specificity, selectivity and safety.

Biomarkers

A biomarker enables the distinction of the cancer cells from that of the surrounding normal cells in the tissue. Cancer pathogenesis as discussed above has been evidenced as a multistage process and further there are a number of different pathways leading to the development of malignancies. Progression to cancer always requires loss of function of genes regulating tumor suppression, DNA repair, and apoptosis apart from gain in function of other genes that facilitate tumor growth, its angiogenesis etc. Knudson's Two Hit Hypothesis [347] states that complete loss of function normally requires that there is a change in both copies of the gene. This could be as a consequence of one of two pathways: 1) acquisition of changes in sequence of either nuclear or mitochondrial DNA. 2) through epigenetic change that is heritable but not sequence related. Such changes result in qualitative and quantitative changes in gene expression and protein production.

The presence of abnormally high levels of such over expressed proteins in the serum or the presence of proteins not normally in contact with the immune system, can elicit a detectable humoral response which may be useful as a biomarker. Thus mutations in the genome either nuclear or mitochondrial, epigenetic changes involving methylation, chromatin structure expression, levels of RNA and protein production can all be biomarkers. Earlier such approaches were designed to compare one or at most a few genes at a time, but today many high throughput technologies have been developed that make it possible to analyze thousands of genes and /or the gene products simultaneously. Cancer biomarker development is thus an important and major area of research that institutions and government's world over are investing their resources and teams have been formed for a collaborative work (Table 3). The launch of a novel drug in the market takes more than a billion dollars and 10 to 15 yrs, and for every successful molecule there are a large number of them that are culled away. Similarly for a novel cancer biomarker to be regularly used at the clinic, there exists a meticulous discovery process, its association with particular disease state, followed by its validation, then the development of this marker by a technique that can be effectively translated at the diagnostic lab, again its validation followed by its approval by the appropriate health regulatory bodies, its marketing and last but not the least educating the technical and clinical personnel.

Table 3. List of institutions involved in collaborative cancer research projects

S. No	Government Institutions
1	International Cancer Genome Consortium(ICGC)
2	Human Proteome Organization(HUPO)
3	National Cancer Institute
4	Cancer Genomics of the Kidney consortium(CAGEKID)
5	National Institute for Health Research
6	EuroTARGET (TArgeted therapy in Renal cell cancer:GEnetic and Tumor related biomarkers for response and toxicity)-European collaboration
7	SCOTRRCC (The Scottish Collaboration on Translational Research into Renal Cell Carcinoma)
8	PREDICT Consortium (Personalized RNA Interference to Enhance the Delivery of Individualised Cytotoxic and Targeted therapeutics)-European Collaboration
9	TCGA (The Tumor Cancer Genome Atlas)-US initiative
10	CGP (Cancer Genome Project)
11	Microarray Gene Expression Data Society
12	CLIA (Clinical Laboratory Improvement Amendments)
13	Centre for Devices and Radiological Health of the FDA

A comprehensive understanding of the molecular mechanisms and cellular processes underlying the initiation of cancer, especially focusing on how small changes in only a few regulatory genes or proteins can disrupt a variety of cellular functions is required, to establish them as biomarkers. A major challenge in cancer diagnosis is to establish the exact relationship between cancer biomarkers and the clinical pathology, as well as, to be able to non invasively detect tumors at an early stage. Similarly, identification of subtle changes in the genomics and proteomics status specific to malignant transformation will allow molecular targets to be used for developing therapeutics. A brief account of the biomarkers reported, ones that are currently employed in clinical oncology for diagnosis and therapy as well as potential ones that particularly hold promise as targets for therapy are compiled here. Thus though many factors may be identified at each step in the progression of the disease, its treatment, impact & prognosis, only a few may be robust enough to be applied in the clinic at the end of all this. In order to provide a quick overview a list of such markers have been compiled here, along with often used techniques and samples, though it is not all inclusive(Table 4). Application of biomarkers in the clinical practice is likely to result in advanced knowledge leading to a better understanding of the disease process that will facilitate development of more effective and disease specific drugs with minimal undesired systemic toxicity. The clinical oncologist at the first interaction with a suspected patient, needs to look for diagnostic and prognostic biomarkers that are quantifiable and enable in

(i)   Identifying who is at risk
(ii) Diagnose at an early stage,
(iii) Select the best treatment modality, and
(iv) Monitor response to treatment

These biomarkers exist in many different forms;

• Traditional biomarkers: include those that can be assessed with radiological techniques viz. , mammograms , X rays, etc., and circulating levels of tumor specific (related) antigens for example, prostate-specific antigen (PSA) in histopathology etc.

The discussion of which is beyond the scope of present chapter.

• Novel molecular biology derived biomarkers which can be categorised as mentioned below are the focus of present R&D efforts world over.

Structural Chromosomal Instability

The presence of chromosomal aberrations is a clear indicator of errors in the DNA damage and mitotic/cell cycle checkpoints. The majority of structural changes as well as specialized classes of chromosomal alterations can only be detected by detailed karyotype analyses. Obtaining sufficient metaphase preparations of adequate quality is the limiting factor. Interphase cytogenetics, using centromere -specific probes by FISH is extremely useful in inferring numerical changes affecting ploidy or leading to polysomy, but cannot determine levels of structural rearrangement. This would lead to a possible misclassification as possessing low or no CIN. The newer molecular cytogenetic assays greatly improve the ability to determine the extent of structural variation of chromosomes which may often be independent of copy number changes; thus enabling investigators to also distinguish a rate of structural chromosomal instability. (S)-CIN can be assessed through the enumeration of the total number of chromosomal breakage and reunion events (i. e. translocation, inversion, deletion or duplication sites) per cell or across a series of specimens made possible through the use of M-FISH/SKY-based analysis, and M- Banding techniques, in the same fashion that interphase FISH using centromeric probes is typically used for the enumeration of copy number CIN [267]. aCGH analysis, can also infer levels of structural genomic complexity, which may be confirmed by parallel metaphase and interphase FISH techniques. The assessment of (S)-CIN, therefore, allows the investigator to integrate the presence and extent of complex structures such as HSRs, DMs, multi-radial chromosomes, rings and ladder-amplification structures within the context of CIN that is independent of gross copy number and ploidy changes. The finding that a stable copy number diploid or aneuploid tumor cell population possesses significantly different levels of (S)-CIN events suggests that molecular pathways impacting repair/checkpoints/cell survival may be fundamentally different to those tumors in which classical CIN predominates. These observations may have an important bearing on the rate of progression, as well as the tumor's metastatic potential and therapeutic response. It can be envisioned that the form of instability, whether copy number CIN or structural, i. e. (S)-CIN, could lead to the predominance of one pathway over another, and it hence follows that the target for treatment will differ accordingly. The presence of high frequencies of structural aberrations as enumerated by the number of translocation/rearrangement sites, suggest pathways affecting DNA repair; whereas changes in ploidy or whole chromosomes suggests pathways affecting mitosis. This critical difference may impact effectiveness of treatments such as the use of mitotic spindle inhibitors or DNA-binding agents.

Table 4. Different Types of biomarkers and techniques used to evaluate them

Type of molecule	Extent of involvement	Technique used
DNA	Whole chromosome numerical/structural	Karyotyping, FISH
	Genes/ nucleotide changes	Sequencing, PCR ,ligation specific PCR ,RT PCR
	Mitochondrial mutation	DNA sequencing/PCR
	Microsatellite analysis	PCR on Short tandem repeats(STR)
	LOH -	SNP genotyping by PCR
	DNA Methylation	Methylation specific PCR (MSP)
RNA	Tissue specific m-RNA	RT PCR/Northern/ RNASE protection /in situ hybridisation
	Tumor specific RNA	RT PCR/Northern/ RNASE protection /in situ hybridisation
	Expression Profile	DNA microarray
Protein	Tumor antigen	Biochemistry, ELISA/Luminex, Flow cytometry, Immuno-histochemistry
	Auto antibody	Biochemistry, ELISA/Luminex, Flow cytometry, Immuno-histochemistry
	Protein marker	MALDI,SELDI and protein array
Cellular		Biochemistry, ELISA/Luminex, Flow cytometry Immuno-histochemistry
Viral		Biochemistry, ELISA/Luminex, Flow cytometry, Immuno-histochemistry
Cancer Antigens		Biochemistry, ELISA/Luminex, Flow cytometry, Immuno-histochemistry

Cytogenetic and Cytokinetic Markers

Structural and numerical aberrations in the chromosomes are classical markers of cancer as the association between chromosomal aberrations and neoplastic transformation has been well established. While deviations from diploid chromosome number leading both to hyper-and hypo-diploid as well as aneuploid have been noted in malignant tumors. Sister chromatid exchanges and translocations give rise to structural aberrations that can be easily scored using various banding techniques (Table 5 & 6). Further, double minutes and homogenously stained regions (indicative of gene amplification) are also often observed in malignant cells that can serve as markers. Although, the ploidy changes complement the clinico-pathological findings, a weak association between ploidy, histological and clinical staging has been noted in many tumors. Somatic mutations (in reporter genes, oncogenes and tumor suppressor genes) are promising biomarkers for cancer risk as these can capture genetic events that are associated with malignant transformation. There is growing evidence that specific polymorphism in certain genes are associated with cancer risk [348-353].

Epigenetic Biomarker

Type specific panels for methylation of different genes have been suggested, e. g., GSTP1, RARβ, TIG1 and APC for prostate carcinoma; p16, RASSF1A, FHIT, H-cadherin and RARβ for non small cell lung cancer VHL, p16, p14, APC, RASSF1A and Timp3 for kidney cancer, etc. [297] as their methylation patterns are known to be altered. Hypermethylation markers may be used for the detection of both primary and metastatic or recurrent cancer cases. For example, hypermethylation of p16 promoter in the circulating serum DNA correlate well with recurrent colorectal cancer [354]. Aberrant methylation of the p16Ink4 and MGMT promoters can be detected in DNA from the sputum of patients with squamous cell carcinoma nearly 3 years before clinical diagnosis , while, methylation of p16Ink4, RASSF1A, or PAX5-beta genes appears to be associated with a 15-fold increase in the relative risk for lung cancer. Therefore, it has been suggested that alterations in methylation patterns of groups of genes in sputum samples may be an effective, non invasive approach for identifying smokers at risk of developing lung cancer [355]. Methylation of the O6-methylguanine-DNA methyltransferase (MGMT) gene, which encodes a DNA-repair enzyme has been shown to inhibit the killing of tumor cells by alkylating agents and methylation of the MGMT promoter of malignant glioma appears to be a useful predictor of the responsiveness to alkylating agents as patients with silencing of this gene seem to respond better to therapy (356). Although, research in epigenetic has led to improved survival of patients with certain forms of lymphoma and leukemias through the use of drugs that alter DNA methylation and histone acetylation, proposed methylation markers need further optimization and large scale clinical trial for further validation. The development of therapeutics that reverse epigenetic alterations in cancer cells, along with prognostic and diagnostic assays based on gene methylation patterns are promising new avenues for future improvements in patient care (Table 7).

Genetic biomarkers

A variety of different sequence changes ,including point mutations, translocations ,deletions, amplification of short repeated sequences, (associated with microsatellite instability) are characteristic of many tumor cells, some cancer cells have up to 10^5 changes in DNA sequence. Both nuclear & mitochondrial DNA can be used to identify sequence variation. Mitochondrial DNA is preferred to nuclear DNA because it is preferentially modified by carcinogens; less efficiently repaired resulting in accumulation of a large number of mutations. Also the cell contains a high copy number generating homoplasmic DNA. Laboratories around the world have teamed up in the ICGC. ICGC coordinates large scale genome studies in 500 tumors from each 50 different cancer types and subtypes in both adults and children adding up to 25,000 cancer genomes. In order, to achieve high sensitivity and specificity genomes of both cancer and normal tissues from the same individual are sequenced. Both exomes and all non coding micro RNAs are also sequenced. This effort is expected to reveal a major repertoire of clinically relevant mutations, subtypes and enable the development of novel therapies. Based on these knowledge specific targeted therapies like trastuzumab for ERBB2 positive breast cancers have been developed. Other such potential targets include: PIK3CA, BRAF, NF1, KDR, PIK3R1 genes, some histone methyltransferases, and demethylases [357, 358]. Novel correlations between the presence of mutation in genes PALB2 in familial predisposition to pancreatic cancer and IDH1& IDH2 in gliomas have been identified [359, 360] Genome wide loss of function, genetic screens, and single well small interfering RNA (siRNA) screens have been used to identify biomarkers that can predict responsiveness to clinically relevant therapeutics. Once resistance or sensitizer genes are identified, their expression is correlated with clinical response to the drug of interest using tumor samples from cancer patients treated with drug in question [361]. Inherited factors are known to play a significant role in cancer. Increased risk due to positive family history is about 25-40%in many common cancers. The first report for familial cancer came from the report that RB1 gene was the causative of retinoblastoma in 1986. Since then more than 100 genes have been identified in hereditary cancer syndromes [362, 363]. Multiple polymorphic variants especially SNPs seem to be involved in cancer susceptibility. There are two hypotheses proposed 1. The common disease-common variant (CDCV) and 2. the common–disease–rare variant. CDCV proposes that multiple common variants each leading to only a minor increase in susceptibility, substantially increase the risk for a common disease. By carrying out Genome wide association studies (GWAS) for susceptibility to more than 20 different cancer types have implicated several dozen susceptibility loci. In colorectal cancer 14 different loci, have been identified some of which are associated with TGF beta signalling, a sequence variant at 4p16. 3 confers susceptibility to urinary bladder cancer, a variant on chromosome 8q24 with colorectal cancer. Thus more genes may be identified as more information comes in from GWAS [364-367]. Among other genome based biomarkers, identification of neoplasm from the level of lesion specific transcriptomes (mRNA of cytokeratin-19, EGFR, MUC 1, etc.) in the blood has been successfully employed in certain epithelial tumors [368, 369]. Recently a novel transcriptome marker based on the levels of exon-3 deleted variant isoform of proghrelin (ghrelin is a growth factor involved in prostate cancer cell proliferation) has been developed that aims at reducing the false positives in prostrate and other endocrine cancers [370]. Enhanced cell proliferation is one of the most important hallmarks of cancer, which is easy to identify using a number of histological, biochemical and flow cytometric analysis. Although subjective in nature, histological assessment based on evaluating the number of mitotic cells present within a given sample, is still used as a routine clinical test and even for grading in certain tumors like breast cancer. Flow cytometric analysis of DNA content, which is automated, objective and rapid allowing large number of cells (and samples) to be measured, has been extensively used for the assessment of proliferation status. (table 6 & 7). This complements histological analysis in most cases, besides allowing analysis of clonal and spatial heterogeneity [348]. Identification of S-phase cells (unequivocal marker of proliferation) and analysis of a number of other antigenic determinants of proliferation (PCNA, Ki67, NOR, etc.) have also been used as complementary markers. Information provided by gene expression analysis has a distinct advantage over other assessments of proliferation (viz., more quantitative, objective, and automated) and could form a component of genomic-based clinical diagnostics of cancer (Table 7). Proteins encoded by the mini-chromosome maintenance (MCM) genes have also been proposed as useful markers of proliferation; with high levels of gene expression indicating poor prognosis. All these genes are cell-cycle regulated and are found among the genes associated with proliferation in tumors [371, 372].

Mitochondrial markers Mitochondria typically contain multiple haploid copies of their own genome (16. 5 kb), including most components of transcription, translation, and protein assembly. MtDNA is present at 1000-10,000 copies/cell, and the vast majority of these copies are identical (homoplasmic) at birth. Several mutations in the mtDNA, particularly in the D-loop region have been recently found in breast, colon, oesophageal, endometrial, head and neck, liver, kidney, leukemia, lung, melanoma, oral, prostate, and thyroid cancer [373]. The majority of these somatic mutations are homoplasmic in nature, suggesting that the mutant mtDNA played an active role in tumor formation [374]. By virtue of their clonal nature and high copy numbers in cancer cells, mitochondrial mutations may provide a powerful molecular marker for non-invasive detection of cancer. It may also be useful in early detection, diagnosis, and prognosis of cancer outcome and/or in monitoring response to certain preventive and interventional modalities as well as therapies [375, 376]. Mutated mtDNA has also been detected in the body fluids of cancer patients and indeed is much more abundant than the mutated nuclear p53 DNA [377]. Since the mitochondrial gene expression signatures of transformed cells have now been identified, development of mitochondrial functional proteomics is expected to identify new markers for early detection and risk assessment, as well as targets for therapeutic intervention [374] (table7). Many advanced techniques are currently available or being newly developed for studying the mitochondrial proteome viz. IP, DiGE, ICAT, SELDI, MALDITOF, protein/antibody array, etc., are expected to facilitate this process further.

Single Nucleotide Polymorphism (SNP Genotyping)

Single nucleotide changes are the most common alterations that are observed in DNA. This change when observed in 1% of the population it is called a SNP if it less frequently observed then it is considered a mutation. The human genome project has identified ultra high density SNP maps. Some of these SNPs could serve as cancer biomarkers as these SNPs are near the genes associated with cancer risk and are in linkage disequilibrium. Linkage analysis and positional cloning studies have identified a number of such genes associated with various types of cancer e. g. BRCA1&2 in breast cancer, colon cancer (FAP familial adenomatous polyposis), and hereditary non-polyposis colorectal cancer (HNPCC) [378]. More recent applications of SNP genotyping involve the global identification of novel LOH (loss of heterogeneity and MSI (microsatellite instability). Each LOH is represented by 20-50 SNPs because of this redundancy, the need to genotype matched normal DNA is eliminated [379]. Detection of SNPs can be done by 2 methods: 1) sequence specific 2) Sequence non specific approach. Sequence specific approach: can be carried out by hybridisation, primer extension, ligation or cleavage. Dynamic allele Specific hybridization (DASH):/SBE: sequence non specific approach relies on the fact that mismatched hetroduplexes of allelic or single stranded DNA have changes in mobility during electrophoresis or liquid chromatography. Thus mobility shift in electrophoresis and HPLC methods are used to detect mismatched hetroduplexes.

Circulating miRNA's. MiRNAs are non-coding, single-stranded RNAs of approximately 22 nucleotides and constituted a novel class of gene regulators, that are differentially expressed in tissues [380], and associated with oncogenesis [381, 382]. In the past few years, there are several reports indicating important aspects of miRNAs in tumor cell lines and tissues. The potential use of miRNAs was focused as diagnostic biomarkers of human cancer in serum-based screening. It is reported that miRNAs could be an ideal class of blood-based biomarkers for cancer detection because: (i) miRNA expression is frequently dysregulated in cancer (ii) expression patterns of miRNAs in human cancer appear to be tissue-specific and (iii) miRNAs have unusually high stability in formalin-fixed tissues [383-387]. This third point led to the speculation that miRNAs may have exceptional stability in plasma and therefore being promising biomarkers for diagnosing human cancers [388-416]. Expression profiling studies have identified that the tissue expression of miRNA can be differentially regulated in human liver diseases and in diverse pathophysiological conditions affecting the liver. There is a need to identify effective biomarkers for diagnosis, prognosis and prediction of treatment efficacy for many liver diseases such as hepatocellular cancer, and chronic viral hepatitis. The identification of disease specific alterations in micro RNA expression and the ability to detect micro RNAs in the circulation provide the basis for identifying novel clinically effective treatments and biomarkers. In hepatocellular carcinoma i. e. primary malignancy of liver and cholangio carcinoma i. e. malignancies arising from biliary tract epithelia, several miRNAs are up regulated or down regulated ,for e.g., :in hepatocellular carcinoma, miR-21, miR-135a, miR-146a, miR-151,miR-221, miR-222 are up regulated and let-7g, miR-22 are down regulated. In Cholangio carcinoma, miR-21, miR-25, miR-31, miR-223, miR-421 are up regulated and miR-122, miR-145, miR-200c, miR-221, miR-222, miR-370, miR-373, miR-494 are down regulated. The unique patterns of disordered miRNA expression in each type of cancer, their stability in serum [417] and their role as biomarkers of disease risk due to inherited polymorphisms suggest that miRNAs may potentially serve as novel molecular biomarkers for clinical cancer diagnosis. Tumor-derived miRNAs such as miR-155, miR-21, miR-15b, miR-16 and miR-24 are detected in the plasmas and serums of tumor patients [418, 419]. These might be a new class of effective biomarkers, and one expects that the miRNAs abundance profile in plasma might reflect physiological and/or pathological conditions. They have revealed that miR-638 is stably present in human plasmas, and miR-92a dramatically decreased in the plasmas of acute leukemia patients. Especially, the ratio of miR-92a/miR-638 in plasma was very useful for distinguishing leukemia patients from healthy persons. Thirteen miRNAs associated with prognostic factors were firstly revealed in chronic lymphocytic leukemia [420]. Among them, miR-15a and miR- 16-1 have been clearly demonstrated to function as tumor suppressors. MiR-15 and miR-16 are located on chromosome 13q14; a region deleted in more than half of B cell chronic lymphocytic leukemia patients, and was found either absent or down-regulated in the majority (~68%) of chronic lymphocytic leukemia patients [381]. Few studies have correlated miRNA expression changes with prognosis in acute myeloid leukemia. AML represents more or less third of young acute myeloid leukemia patients. The prognostic signature included miR-181a and miR-181b which were inversely correlated with risk of event and miR-124, miR- 128-1, miR-194, miR-219-5p, miR-220a, miR-320 which were positively associated with the risk of event. But no conclusive evidence of their circulation in blood has been reported. High levels of BIC and mir-155 were detected in Hodgkin lymphoma. Tumor-associated miR-155, miR-210 and miR-21 were found in serum of diffuse large B-cell lymphoma (Hodgkin lymphoma) patients and healthy controls [418]. They showed that circulating miRNAs were clearly detectable in serum samples and that higher levels of specific miRNAs were associated with diagnosis and prognostic outcome in Hodgkin lymphoma patients. MiRNAs are highly significant not only in physiological processes but also in pathological processes and tumorogenesis. To date, miRNAs expression profiles in many types of cancers have been identified and miRNAs expression signatures associated with types of leukemia have also been reported. A large number of researchers have found that significant differences were presented between the exosomal miRNA levels for the lung adenocarcinoma group and the control group [421-423]. MiR-21 and miR-155 were elevated in lung cancer . High expression of miR-155 has correlated with significantly shorter survival, and low expression of let-7a-2 has conferred poor prognosis in resurrected lung adenocarcinoma. Intriguingly, in 4 lung adenocarcinoma cases in which paired tumor and plasma samples were examined, there was a close correlation between circulating miRNAs of tumor-derived exosomes and tumor miRNAs, confirming that miRNA expression in peripheral blood could be a surrogate of miRNA expression in the tumor biopsy. Thereafter, several reports show that exosomes could be an important resource of cell-free miRNA in serum or plasma [424, 425]. The expression of specific circulating miRNAs is a good surrogate marker of tumor. In ovarian cancer higher expression of miR-200 was associated with poor prognosis [426]. Let-7i might be used as a therapeutic target to modulate platinum-based chemotherapy and as a biomarker to predict chemotherapy response and survival in patients with ovarian cancer. MiR-9 and miR-223 can be of potential importance as biomarkers in recurrent ovarian cancer [427]. Resnick et al [428] used 28 serums of epithelial ovarian cancer patients to determine the utility of serum miRNAs as biomarkers for diagnosis. MiRNAs-21, 92, 93, 126 and 29a were significantly over expressed in the serum from cancer patients compared to controls. MiRNAs-155, 127 and 99b were significantly under-expressed. MiR-21 may serve as a molecular diagnostic & prognostic marker for breast cancer and disease progression. Recently, using quantitative real-time polymerase chain reaction (RT-PCR), the expression levels of miRNAs in plasma from a series of colorectal cancer patients and controls [429] were examined. It was found that the expression of miR-92 in plasma could distinguish colorectal cancer patients from healthy control patients with 70% specificity and 89% sensitivity. MiRNAs contribute to carcinogenesis and may provide new therapeutic strategies for cancer, but it is too early to know if expression levels of circulating miRNAs will be of sufficient diagnostic value for their use as a screening method for colorectal cancer. Expression profiles of miRNA in oesophageal adenocarcinoma were first reported. [430, 431]. MiRNA expression of 35 frozen specimens, 10 adenocarcinomas, 10 squamous cell carcinomas, 9 normal epithelium, 5 Barrett's oesophagus and 1 high grade dysplasia) was analyzed using Ambion bioarrays containing 328 human miRNA probes. They found that mir-203 and mir-205 was expressed 2-10 fold lower in squamous cell carcinomas and adenocarcinomas compared with normal epithelium. Mir-21 expression was 3-5 folds higher in both tumors versus normal epithelium. In 2009, Mathe found miR-21, miR-223, miR-192, and miR-194 expression was elevated, whereas miR-203 expression was reduced in cancerous compared with noncancerous tissue in adenocarcinoma patients [432]. A set of 4 human miRNAs (miR-28, miR-185, miR-27, and let-7f-2) in 245 miRNAs were found significantly up regulated in renal cell carcinoma compared to normal kidney.

Therefore, based on these researches, circulating mi RNAs are promising future application for diagnosing human cancers. Still, large, independent, well-characterized, family and population-based case control and additional validation studies are warranted to determine whether circulating miRNAs could serve as biomarkers of human cancers

Cells as Biomarkers

CTC

Cellular biomarkers are also called as Circulating Tumor Cells (CTC). Tumor cells appear in the blood stream in advanced stages of cancer. The presence of CTC's at multiple time points throughout the course of therapy enables prediction of survival in patients with metastatic breast cancer. Elimination of CTC indicates effectiveness of therapy. CTCs may not only be surrogate end points but may also aid the selection of candidates for clinical trial. They have been reported to be superior to standard tumor markers [433-437].

T regulatory Cells (CD4+, CD25+and Foxp3+)

The immune system has a number of mechanisms that contribute to the ability to discriminate self from nonself. Regulatory T cells (T regs) are reported to be important in inducing and maintaining peripheral self tolerance and thus preventing immune pathologies. [438, 439]. These are sub population of CD4 expressing high levels of CD25 and FoxP3. Increased levels of T regs have been reported in: lung, pancreatic, breast, liver and skin cancers either in serum or the tumor itself [440-442]. In ovarian carcinoma patients, presence of Tregs which suppress tumor specific T cell immunity correlates inversely with survival (443). Tregs may serve as a surrogate immune marker of cancer progression and predictor of therapeutic response as well. The presence of FoxP3$^+$ cells in tumors has been reported to predict prognosis, invasiveness and metastatic ability of some tumors by modulating the ability of the immune system to target tumor cells [444], thus Treg and Fox P3 should be explored as biomarkers as it can also be detected in peripheral blood thus its application to laboratory medicine becomes more promising.

Cancer Stem Cells

The cancer stem cell model for tumorogenesis hypothesisizes that within the tumor there is a sub population of cells that have the capacity to self renew thereby enabling differentiation and maintenance of the tumor and these are called Cancer Stem cells (CSCs). The presences of CSCs were first reported in AML and subsequently in glioblastoma, medulloblastoma, breast cancer, melanoma and prostate cancer. It appears that identifying CSCs for every possible tumor is important to include it as a biomarker. It also leads to novel therapeutic avenues [445, 446]

Viral

The viruses well known for their association with particular tumor are: Hep B, Hep C with hepato-cellular carcinoma, human papilloma virus, cervical and genital warts, ,Epstein Barr virus with Burkitt's lymphoma, Hodgkin's & non Hodgkin's lymphoma, nasopharyngeal carcinoma, lymphomas ,leiomyelosarcomas, HIV patients with EBV virus. Adenovirus and other retroviruses are also associated with cancers. These viruses integrate in the host cell genome so they are perpetuated. The integration of viral oncogene (v oncogene) at the cellular oncogene (c- oncogene) has long been implicated in malignant transformation. Detection and quantification of these viral nucleic acids or their gene products in serum /plasma can serve as potential diagnostic as well as prognostic markers and for staging as well. They also are good for monitoring therapeutic response [447-450].

Cancer Antigens Based Biomarkers

The cancer proteome contains information on perhaps every biological process that takes place in cancer cells, cancer tissue microenvironment, and cancer cell-host interaction. Cancer cells release many proteins and other macromolecules into the extra-cellular fluid through secretion that can also serve as biomarkers. Some of these products can end up in the bloodstream and hence serve as potential serum biomarkers. Some important cancer antigens that serve as diagnostic and prognostic biomarkers of cancer are summarized in the Table 6 & 7.

Prostate Specific Antigen (PSA)

Prostate specific antigen belongs to the serine protease family of "Kallikrein genes "and hK2 & hK3 expression is restricted to prostrate in males. It is expressed both in normal and cancerous conditions as well. It appears to be involved in the initiation and growth of prostate cancer by abnormal release of growth factors or proteolysis of growth factor binding proteins. Also known to have a role in invasion and metastases through the degradation of collagen and laminin [451]. PSA is present in small quantities in the serum of normal men, and is often elevated in the presence of prostate cancer and other prostate disorders. However, prostate cancer can also be present in the complete absence of an elevated PSA level [452]. Therefore it may also be associated with other prostrate related conditions. A blood test to measure PSA is considered the most effective test currently available for the early detection of prostate cancer. Though it is an easy test it may not serve as a decisive biomarker other clinical correlates are required. (Table 6).

Alpha-fetoprotein (AFP) is the major serum fetal protein in mammals, which is actively produced and secreted during the fetal life by the liver hepatocyte. It is a common biomarker used in hepato-cellular carcinoma, but is not sensitive or specific enough hence cannot be used as biomarker in itself. Since the levels of AFP may be elevated in serum from patients with other chronic liver disease; AFP is not useful for screening in patients suffering from liver cirrhosis or hepatitis C. It is also secreted in neuroblastoma, hepatoblastoma, and hepatocellular carcinoma [453-455].

CA125: The CA 125 antigen is a membrane glycoprotein produced by tissues derived from coelomic epithelium that is expressed by most epithelial ovarian cancers. CA125 was initially detected in 1983 using the monoclonal antibody designated OC125; hence the name CA125. CA125 is a powerful index of risk of ovarian and fallopian tube cancer in asymptomatic postmenopausal women. It is found in the serum of more than 80 per cent of the patients with epithelial ovarian tumors, with half life of 4 days. CA-125 antigen remains the only serum tumor marker routinely used in epithelial cancer of the ovary for patient prognosis, disease progression, and response to chemotherapy. CA125 is also expressed by a number of tissues of both cancerous and noncancerous origin. It may also be elevated in other malignant cancers, including those originating in the endometrium, fallopian tubes, lungs, breast and gastrointestinal tract. Certain physiological conditions also modulate CA125 levels, as the levels are elevated slightly during menstruation and more prominently during the first trimester of pregnancy. A decrease in CA125 level is generally associated with tumor response to therapy, whereas a rising level is suggestive of drug resistance. Indeed, CA125 is an accurate marker to define relapse of ovarian cancer. CA125 has numerous applications in the design of clinical trials, from prognosis to follow up and is a tool that is complementary to standard criteria for disease measurements. The key problems in using the CA125 test as a screening tool are its lack of sensitivity and its inability to detect early stage cancers [456-461].

CA15-3: The CA15-3 protein is a member of the family of proteins known as mucins, whose normal function is cell protection and lubrication. It plays a role in reducing cell adhesion and is found throughout the body. Elevated levels of this antigen are found mainly in breast cancer where it appears to be involved in metastasis. Patients with other cancers such as lung, colorectal, ovarian, pancreatic and hepatic dysfunction also have elevated levels of CA-15-3. In breast cancer it is been used to monitor disease progression, metastasis, treatment response [462, 438, 463].

CA 19-9 (cancer antigen 19-9) or GICA (gastrointestinal cancer antigen) is a glycolipid with unknown biological function, which was the first successful tumor marker used for serological diagnosis of pancreatic cancer. Reported to be a sensitive and specific marker in serum for pancreatic cancer, its elevated levels in urine have been found in bladder cancer, Urothelial dysplasia or carcinoma in situ. Owing to its high specificity, it plays an important role in the diagnosis, therapeutic monitoring and monitoring of the course of gastrointestinal carcinomas, in particular in the case of pancreatic carcinoma, hepato biliary carcinoma (carcinoma of the liver, carcinoma of the bile ducts) and carcinoma of the stomach. Patients with non malignant inflammatory diseases, such as cholecystitis and obstructive icterus, cholelithiasis, cholecystolithiasis, acute chlolangitis, toxic hepatitis and other liver diseases also have increased levels of CA 19-9 therefore should be used with caution [464-467].

Carcinoembryonic antigen (CEA) is a 200 kDa glycoprotein, discovered by raising antibody against extract of human colonic carcinoma. Elevated levels are found in patients with colorectal, breast, lung, or pancreatic cancer, and also in smokers. The first success in developing a blood test for CEA was in 1965, when the antigen was detected in the blood of some patients with colon cancer. Blood levels of CEA are also elevated in many other cancers such as those of the thyroid, pancreas, liver, stomach, prostate, ovary, and bladder. Post-operative normalization of serum CEA level has been reported to be a favourable prognostic indicator in lung cancer and the identification of abnormal pre- and post-operative serum CEA levels may be useful in the auxiliary cancer prognosis or post-operative surveillance of colorectal cancer patients [468-470].

Human chorionic gonadotrophin (HCG) is a hormone produced normally by the placenta, whose level is elevated in the blood of patients with certain types of testicular and ovarian cancers (germ cell tumors) and choriocarcinoma. In addition, synthesis of free βhCG and its subunits by pelvic carcinomas such as those of the colon, urinary tract, prostate, uterus and vulvo-vagina have also been reported. The presence of increased serum levels of hCG and its metabolites is generally considered to be a sign of a poor prognosis and it has been suggested that βhCG might directly modify the growth of the cancer, leading to a worse outcome. The clinical use of free βhCG as a tumor marker has been limited to a small number of patients owing to a short half life and rapid renal clearance. An elevated blood level of HCG is also found in the urine of pregnant women and therefore may not be useful as a marker under this condition [471,472].

Thyroglobulin (Tg) is a large glycoprotein stored in the follicular colloid of thyroid gland and that acts as pro-hormone in the intra-thyroid synthesis of thyroxin (T4) and triiodothyronine (T3). It is an organ-specific tumor marker; associated mainly with patients harbouring differentiated thyroid cancer that arise from the follicle cells, often resulting in increased levels of thyroglobulin in the blood.

It is used as a tumor marker to evaluate the effectiveness of treatment for thyroid cancer and to monitor recurrence [473]. Tg is generally measured in serum, but measurements can also be made in thyroid cyst fluids and other fluids/tissue obtained by fine needle biopsy of thyroid nodules [474]. Concomitant measurement of anti thyroglobulin is essential as anti-Tg antibodies interfere with the Tg assay and render the assay for thyroglobulin invalid [475]. This important factor should be taken into consideration during the evaluation of patient's status.

Further, this will not be a very useful marker in tumors which do not release a significant amount of thyroglobulin into the circulation. Although an undetectable level of serum Tg after thyroidectomy and I^{131} ablation suggests that patients are free of disease, several studies have shown that a minority of patients, under the influence of TSH stimulation, have elevated serum Tg levels and therefore interpretation needs to be made with caution [476].

Heat shock proteins (HSPs) are proteins expressed in response to stress, each kind of stress having its unique protein [477]. Heat shock proteins (Hsps) are over expressed in a wide range of human cancers and are implicated in tumor cell proliferation, differentiation, invasion, metastasis, death, and recognition by the immune system. At present it is unclear as to how the Hsps become over expressed in cancer; one hypothesis is that the physio-pathological features of the tumor microenvironment (low glucose, pH, and oxygen) stimulate the Hsp induction [478, 479].

Although Hsp levels are not informative at the diagnostic level, due to over expression in a wide range of malignant cells and tissues, these are useful biomarkers for carcinogenesis in tissues and are suggestive of the degree of differentiation and the aggressiveness in certain types of cancers. Further, the circulating levels of Hsp and anti-Hsp antibodies in cancer patients may be useful in tumor diagnosis.

Several Hsp are implicated in the prognosis of specific cancers; most notably Hsp27, whose expression is associated with poor prognosis in gastric, liver, and prostate carcinoma, and osteosarcoma, while Hsp70 is correlated with poor prognosis in breast, endometrial, uterine, cervical, and bladder carcinomas. Increased Hsp expression has also been found to predict the response to certain anticancer treatments. While Hsp27 is associated with a poor response to chemotherapy in leukemia patients, Hsp70 expression predicts a better response to chemotherapy in osteosarcomas. Implication of Hsp in tumor progression and response to therapy has led to its successful targeting in therapy by use of Hsps in anticancer vaccines, exploiting their ability to act as immunological adjuvant [480].

Other proteins Oral fluid contains proteomic signatures that may serve as biomarkers for human diseases such as oral cancer. Therefore, it has been suggested that proteomic analysis of human oral fluid such as whole saliva holds promise as a non-invasive method to identify biomarkers for human oral cancer [481]. Most recently, detection of five proteins in the saliva of cancer patients has been found to be useful markers of oral cancer with 90 per cent sensitivity and 83 per cent specificity for oral squamous cell carcinoma. These proteins include (i) calcium-binding protein MRP14 implicated in several types of cancer; (ii) CD59 over expressed on tumor cells that enables them to escape from complement-dependent and antibody-mediated immune responses; (iii) Profilin 1, a protein involved in several signalling pathways with cytoplasmic and nuclear ligands, generally secreted into tumor microenvironments during the early progressive stage of tumorogenesis; and (iv) catalase, a member of the enzymatic anti-oxidative system, whose level is elevated in many human tumors and involved in carcinogenesis and tumor progression [482]. However, long-term studies employing large number of oral cancer patients as well as subjects at high risk of developing oral cancer are needed to validate these potential biomarkers.

Cancer Associated Serum Antibodies

During the process of malignant transformation a number of tumor specific antibodies are expressed and released in the serum. These proteins being non cellular are present in high levels are rendered immunogenic and therefore give rise to an immune response, resulting in expression of circulating antibodies, these are identified as a class of auto-antibodies. Such antibodies can be easily detected with assays routinely carried out in most laboratories. Also these antibodies may also be present before the clinical diagnosis of cancer thus making it a good candidate biomarker. Though promising none yet have been established with adequate sensitivity [483-487].

Biomarker Validation

Cancer biomarkers has been an area where so much data has accumulated in so short a time. The enabling technologies in identifying DNA, RNA and protein with all their variations however gross or minute has been so overwhelming in the post "omics" era. that the list of biomarkers identified are long . The list though promising to enable effective, early diagnosis, treatment and prognosis is yet to translate to the clinic. This is because the validation of these markers is more difficult, time consuming and cost intensive than the discovery itself.

The validation may require multi centric studies across the globe and processes are now in place to enable this, it has been a concerted effort between institutions and nations. The major challenges are: Tumor heterogeneity and global profiling approaches. The facilitating factors would be that, all stake holders come to consensus on carrying out collaborative work, ensure standardisation and reproducibility, address experimental variability, have standardised protocols for experimentation, data acquisition, normalisation, reducing inter-laboratory variation. Thus all this would take its slow course though eventually it will happen!!.

Clinically Relevant Diagnostics

The intent to improve and alleviate suffering, to offer early effective diagnosis, therapy and disease management is the goal of all the stakeholders of healthcare. Research is a dynamic process offering incremental advantages of magnitudes of higher order to, pre-existing technologies. Despite the breakthrough technologies and understanding of the disease progress the immediate task at hand is to enquire what among these are relevant biomarkers, what kind of sample is it be obtained from the patient now and today at the clinic.

Most often these measurements are made on serum as blood, and urine are amongst the cheapest easiest samples that can be obtained with considerable ease. However more expensive and invasive methods are also used for diagnosis such as: good quality sputum, fine needle biopsy, tissue biopsy, bone marrow, lung exudates, pleural fluid, body fluids, and chorionic villi. Traditional biomarkers fall into five categories: Protein, biochemical, histopathology, cytogenetic and imaging methods.

Proteins: Serum samples are typically used to detect levels of enzymes, hormones and other proteins such as cytokines immuno-globulins etc. These assays are relatively easy, rapid and convenient assays used for long time in the clinical set up. The levels of these in the blood may indicate damaged diseased state, or organ dysfunction. Serum tumor markers are particularly useful for early detection, prognosis, and tumor staging. Some serum tumor markers are unique to specific cancers such as CBH Ag 125, and PSA for ovarian and prostate cancer respectively, while carcinoembryonic antigen is more general indicator for colorectal, lung, ovarian, breast, pancreatic, and gastrointestinal cancers [488, 489]. Biomarkers used are listed in Table 6.

Chemical Tests

The basic metabolic profile commonly referred to as CHEM 7 is the basic hallmark of modern medicine. The sample material for these tests is blood /urine. This read out provides general information on the patients metabolic profile which is then correlated with the clinical symptoms and if required more specific tests are carried out such as kidney function, lipid profile, (total cholesterol, HDL, LDL, triglycerides,) homocysteine specific disease risk information. More recent markers are listed in Table 7.

Cytogenetic

Sample material: peripheral lymphocytes from whole blood, bone marrow, amniotic fluid chorionic villi and sometimes even tissue samples from solid tumors. Cells from these sources are amplified as primary cultures, cells are arrested at metaphase or late prophase by addition of colchicine a spindle inhibitor, subjected to hypotonic treatment, and fixed in Carnoy's fixative and cast on microscopic slides. The cells are stained by GTP banding, and other banding techniques and metaphase chromosomes are screened for gross numerical and structural anomalies using various commercially available softwares. However minute aberrations are not captured by this method, hence more recent techniques like FISH and its modifications are used. The fluorescent probes are used to detect suspected regions using region specific probes or whole chromosome probes are used when prior knowledge is not available . This technique is now routinely applied in diagnostic labs. Various anomalies associated with different conditions are listed in Table 5 & 6.

Though Philadelphia chromosome i. e. translocation 9:22 /Bcr–Abl was the first documented and even today a routine diagnostic for CML. , presently more than 500 recurrent chromosomal rearrangements in cancer have been identified. More than 275 chromosomal break points have been characterized. Most of these are in hematological anomalies; however some validations are there in solid tumors. The problem with this to obtain good sample at the time of biopsy and clonal heterogeneity causes less accurate enumeration and therefore dependability. They may be however correlated with supportive diagnostic tests.

Table 5. Leukemia Biomarkers used for Prognosis and Diagnosis

Type of cancer	Abnormality(s)/Genes(s)	Biomarker	Application	Sample/ Detection Method
Hematological Tumors				
Leukemia				
AML	1)MLL gene on 11q23 with different partners: 4q21,19p13. 3,9p22,22q13,1q21 2)TEL gene on 12p13 with different partners: 5q33,9q34,22q11 3)inv(3)(q21q26)/ROPN1;EVll 4) -5/? 5) t(6;9)(p23;q34)/DEK;CAN 6) -7/? 7) -5,del(5q),-7 and/or+8/? 8)del/t(11q23)MLL(ALL-1)	Genetic Biomarkers Cancer stem cells	D D,P&T	PBL or BM/FISH Tumor sample/ Immuno- cytometry
AML M5b	9)t(12;21),ETV6-RUNX1(TEL-AML1)			
AML M2	t(8;16)(p11;p13)/MYST3 (MOZ);CREBBP(CBP)			
AMLM 5	t(8;21)(q22;q22)(CBFA2T1 (ETO);RUNX2(CBFA1),RU NX1(AML1)			
AML M3	t(9;11)(p21;q23)/MLLT3 (AF9);MLL(ALL-1)			
AML M4EO	t(15;17)(q22;q11)/NPM; RARA inv(16)(p13q22)/MYH11;CB FB2			

Type of cancer	Abnormality(s)/Genes(s)	Biomarker	Appli-cation	Sample/ Detection Method
APL	1)RAR-alpha on 17q21 with different partners:15q21,11q23 2)RARA-PML mutation 3)RARβ2 gene	Genetic Biomarkers	D	PBL or BM/FISH/P CR
ALL	1)C-MYC on 8q34 with different partners: 14q32,2p12,22q11 2)TCR-alpha/delta on 14q11 with different partners:8q24,11p15,11p13 3)TCR-Beta on 7q34 with different partners:1p32,9q34,19p13 4)t(4;11)(q21;23)/MLLT2(AF 4);MLL(ALL-1) 5)t(5;14)(q31;q32)/IL3;IGH@ 6)t(9;22)(q34;q11)/ABL;BCR	Genetic Biomarkers	D	PBL or BM/FISH
TPLL	TCR Alpha/delta on 14q11 with different partners:14q32. 1,Xq28	Genetic Biomarkers	D	PBLor BM/Cyto-genetics
CML	1)t(9;22)(q34;q11) t(9;12)(q34;p13) 2)BCR-ABL translocation	Genetic Biomarkers	D	PBL or BM/Cytoge netics/FISH /PCR
MDS	1)t(3;21)(q26;q22),t(12;22)(p1 3;q1) 2)t(7;11)(p15;p15) del(5q)/?	Genetic Biomarkers	D	PBL or BM/FISH
CLL	1)del(11q22)/? 2)del(13q14)/? 3)del(17p13)/TP53(cr44) 4)t(15;17)PML RARA,t(8;21)AML1-RUNX1T1(AML-ETO),inv(16)	Genetic Biomarkers	D	PBL or BM/FISH

Non-Hodgkin's Lymphoma				
B cell Lymphoma				
Lympho plasma-cytoid lymp-homa	t(9;14)(p13;q32)	Genetic Biomarkers	D	Lymph node or BM or biopsy/IHC/ Flow Cytometer PCR/FISH/ Paraffin blocks
Folli-cular lympho ma	1)t(9;14)(p13;q32)-(CR21) 2)t(14;18)(q32;q23)/BCL2;IGH @	Genetic Biomarkers	D	PBL or LA/FISH
MALT	1)t(11;18)(q21;q21),MALTI gene 2)t(1;14)(p22;q32)	Genetic Biomarkers	D	PBL or LA/FISH
Mantle cell Lym-phoma	1)t(11;14)(q13;q32)Cyclin D1 and IGH genes	Genetic biomarker	D	PBL or LA/FISH
Bur-kitt's Lym-phoma	1)t(8;14)(q24;q32),t(2;8)(p12;q 24) 2)t(8;22)(q24;q11) 3)c-myc translocation	Genetic Biomarkers	D	PBL or LA/FISH
DLBCL	1)t(14;18)(q32;q21) 2)t(3;14)(q27;q11) 3)3q27 aberrationsc-myc mutation	Genetic Biomarker	D	LA/ PCR/FISH
T-cell Lymphoma				
T cell anapla-stic large cell lymp-homa	1)t(2;5)(p23;q35)- 2)t(1;2)(p21;p23) 3)t(2;3),(p23;p20) 4)t(2;22)(p23;q11) 5)inv(2)(p23;q35)	Genetic Biomarker	D	PBL or lymph/FISH

Table 6. Solid Tumor Biomarkers for Prognosis and Diagnosis

Cancer Type	Abnormality(s)/Gene(s)	Biomarkers	Appli cation	Sample type/Metho d of Detection
Glial Tumors	del (1p) and/or del(19q)/?	Genetic Biomarkers	D	Formalin fixed or biopsy/FIS H/RTPCR
Alveolar rhabdomyosarc oma	1)t(2;13)(q35;q14)/PAX3;F OXO1A) 2)t(2;13)(q35;q4) 3)t(1;13)(p36;q14)	Genetic Biomarkers	D	Frozen sample or Formalin fixed sample/FIS H/Cytogene tics/PCR
Salivary gland	1)t(3;8)(p21;q12)/? SMAD4(TGF-β)	Genetic Biomarkers	D	Biopsy/Cyto genetics/FIS H/Spectral
Pleomorphic adenoma	Rearrangements of 3p21,8q12 and 12q15			karyotyping /PCR/IHC
Adenocarcino ma	del(6q)			
Adenoid cyctic carcinoma	t(6;9)(q21-23;p13-22)			
Mucoepidermo idarcinoma	del(6q),t(11;19)(q14-21;p12-13)			
Ovary	1)t(6;14)(q21;q24)/? BRCA1,BRCA2 2)p73(Imprinted genes) 3)PTEN	Cancer antigen 125(CA125)	D&P	Serum/imm unoassay
		Human chorionic gonadotropin (hCG)	D	Serum/ELIS A
		APC gene	D&P	Blood,tumo rsample/RF LP of chromosom e 5q21-22, methylation status of APC gene
	4)PIK3CA amplification	Genetic Biomarkers	D	FISH

Thyroid	1)RET 2)PRKAR1A(Carney complex), 3)PTEN/MMAC1/TEP1(Cowden)	Genetic Biomarker	D	PCR Blood,tumorsample/RFLPofchromosome5q21-22,methylation status of APC gene-
	4)APC	APC gene	D&P	
Thyroid adenoma	+5,+7,+12			
Thyroid papillary carcinoma	1)inv(10)(q11q21) 2)t(7;10)(q35;q21)/?	Thyroglobulin(Tg)	D&P	Serum/ELISA or IHC with TPO Ab
Astrocytic tumor	1)amp(7p)/EGFR 2)NF1 3)TSC1,TSC2	Genetic Biomarkers	D	Biopsy/Cytogenetics/RFLP/
Skin	1)BRAF mutation 2)t(9;16)(q22;p13)/? CDKN2A,CDK4(Familial melanoma) NF1,NF2 3)FISH assay targeting 6p25,6q23,Cep6 and 11q13/11q14/11q15	Cancer stem cells(CSCs) Genetic Biomarker	D,P&T	Tumorsample/immunocytometry FISH
Ewings Sarcoma	t(11;22)(q24;q12)/FLI1;EWSR1	Genetic Biomarkers	D	PB or Parffin embedded tissue/Nested/RTPCR/PCR
Askins tumor	t(21;22)(q22;q12)		D	Cytogenetics/FISH/
Peripheralneuroepithelioma	t(7;22)(p22;q12)		D	Cytogenetics/FISH
Liposarcoma	amp(12q)/numerous	Genetic Biomarkers	D	Biopsy or FNAC/Nested RTPCR
Liposarcoma,myxoid and round cell	t(12;16)(q13;p11)			
Liposarcoma well differentiated	r(12)			

Breast	1)amp(17q12)/ERBB2- 2)BRCA1,BRCA2,PRKAR 1A(Carney complex), 3)PTEN/MMAC1/TEP1(Co wde) 4)TP53,hCHK2(LI-i- Fraumeni) 5)STK11/LKB1 6)p14^{ARF}(cell cycle) 7)14-3-δ (cell cycle) 8)ER ,PR,RAR- β2,(Hormonal Receptor) 9)E- Cadherin(Invasion,Metastasi) 10)Peg1/MEST(Imprinted genes) 11)ErB2(Receptor tyrosine kinase)	Cancer antigen 15- 3(CA15-3)- BRAC1,BR AC2 Circulating tumor cells(CTCs) Cancer stem cells(CSCs)	D&P D D&P D,P& T	Serum/ELIS A/Lymph node or BM/IHC/ Tumor samples/RT PCR Blood/Imm unocytometr y Tumor sample/imm unocytometr y
	12)HER2,TOP2A,ESR1,CC ND1 amplification- 13)EML4-ALK rearrangements	Genetic Biomarker	D	FISH/IHC
Dermatofibros arcoma	1)t(17;22)(q22;q13)/COL1A 1;PDGFB	Genetic Biomarker	D	Paraffin embedded tissue/FISH
Synovial sarcoma	1)t(X;18)(p11;q11)/SYT;SS X1;SSX2 2)t(X;18)(p11;q11)	Genetic Biomarker	D	Paraffin embedded tissue/FISH/ IHC/
Gastrointestina l stromal tumors(GIST)	c-KIT mutation	Genetic Biomarker	D	Biopsy/IHC
Gastrointestina l cancer	1)APC 2)MADH4/SMAD4/DPCA/ BMPR1A 3)STK11/LKB1			
Pituitary	1)PRKAR1A 2)MENI	Biochemical Marker	D	Whole blood/Imagi ng/Biochem ical analysis/Ser um

Testicular	1)PRKAR1A 2)STK11/LKB1	Human chorionic gonadotropin	D	Serum/ELIS A
Endometrial neoplasms	1)PTEN/MMAC1/TEP1(Co wden) 2)MLH1(DNA Repair) 3)amp(ESR1)	Genetic Biomarkers	D	Biopsy/FIS H
Endometrial adenocarcinom a	1)P27(Cell cycle) 2)hMSH2,hMSH6,hMLH1, hPMS1,hPMS2			
Pancreatic cancer	1)CDKN2A,CDK4(Familial melanoma) 2)MEN1 3)RASSF1(Signal transduction) 4)SMAD4(TGF-β)	Cancer antigen 19-9(CA19-9)- APC gene	D&P D&P	Serum/ELIS A Blood.tumo rsample/RF LP of chromosom e 5q21-22,methylati on status of APC gene
	5)LOH -Retinoblastoma gene	Genetic Biomarker	D	FISH
Renal cell carcinoma	1)MET 2)TSC1,TSC2 3)VHL(CR44) 4)P73(Imprinted genes)- 5)del(3p),der(3p),t(X;1)(p11 ;q21)	Genetic Biomarker	D	Smears/Cyt ogenetics/FI SH
Renal cell adenoma	+7,+17,-Y			
Renal Carcinomas	1)WT(Wilms Tumor) 2)H19(Imprinted genes) 3)TIMP3(invasion,metastasi s			
Paraganglioma	SDHD,SDHC,SDHB	Genetic Biomarker	D	Biopsy/Mult iplex PCR
Pheochromocy toma	1)SDHD,SDHC,SDHB 2)RET 3)VHL	Genetic Biomarker	D	MSSCP,RF LP,Direct sequence analysis/FN AC/Biopsy

Colorectal cancer	1)hMSH2,hMSH6,hMLH1, hPMS1,hPMS2 2)MADH4/SMAD4/DPCA/ BMPR1A 3)APC,LEF1,TCF4,β-catenin(Wnt/TCF) 4)KRAS Mutation		D	Tumor specimen/Tr aditional sequencing/ rtpcr/RFLP/ HRMA(Hig h resolution melting analysis-
	5)EML4-ALK rearrangements 6)PIK3CA amp	Genetic Biomarkers		FISH
		Carcinoembr yonicantigen (CEA)	D&P	Serum/ELIS A
Colorectal adenomatous polyps	APC	Genetic Biomarker	D	PBL/PCR
Sarcomas	TP53,hCHK2(Li-Fraumeni)	Genetic Biomarker	D	Tumor Biopsy or formalin fixed Paraffin embedded/R TPCR
Clear cell sarcoma	t(12;22)(q13;q14)			
Brain tumors	1)TP53,hCHK2(Li-Fraumeni) 2)MGM(Apoptosis)-	Cancer stem cells(CSCs)-	D,P& T	Tumor sample/Imm unocytometr y
Adrenocortical tumors	TP53,hCHK2(Li-Fraumeni)	Biochemical marker	D	FNAC/Biop sy/Biochemi cal analysis
Basal cell carcinomas	RET/PTC	Genetic Biomarker	D	Biopsy/Mult iplex PCR
Meningiomas	NF2	Genetic Biomarker	D	FNAC/Cyto genetics/FIS H
Retinal tumors	RB1 VHL	Genetic Biomarker	D	PBL/RFLP

Colon Cancer	1)MBD4- 2)p14^{ARF}(Cell cycle) 3)APC(Cell cycle) 4)RASSF1(Signal transduction) 5)MGM(Apoptosis) 6)MLH1(DNA repair) 7)PR,RARβ2(Hormonal receptor) 8)TIMP3(Invasion,Metastasis) 9)Peg1/MEST(Imprinted genes) 10)Ras 11)amp(MDM2)	Genetic Biomarker	D	PBL/FISH
Multiple myeloma	1)P15^{INK4b}(cell cycle)(CR5) 2)LOH-TP53	Genetic Biomarker	D	Biopsy/FISH
Gastric cancer	1)Her2,Her4 ,MET amplification 2)p14^{ARF}(cell cycle) 3)14-3δ(cell cycle) 4)MGM(Apoptosis) 5)ECadherin(Invasion,Metastasis)	Genetic Biomarker Heat shock proteins(Hsp 27,Hsp70)	D D&P	IHC/FISH- Serum/ELISA
Esophageal cancer	1)p14^{ARF}(cell cycle) 2)APC,LEF1,TCF4,βcatenin (Wnt/TCF) 3)MET gene amplification	Genetic Biomarker	D	Biopsy/FISH
Hepatocellular carcinoma	1)14-3-δ(cell cycle) 2)ECadherin(Invasion,metastasis)	Alpha-foetoprotein(AFB)	D&P	Serum/imm unoassay
Neuroblastoma	1)amp(2p)MYCN 2)amp(17q21) 3)Caspase-8(Apoptosis)-	Genetic Biomarker	D	Biopsy/CGH,FISH/Multiplex PCR/Flowcytometry/RTPCR
Nasopharyngeal Cancer	1)RASSF1(Signal transduction) 2)DAPK(Apoptosis)	Genetic Biomarker	D	Blood and body fluids/Methylationspecific PCR

Prostate cancer	1)ER 2)PTEN 3)t(TMPRSS2 ;ERGorETV1) 4)t(TMPRSS2 ;ERGorETV2) 5)t(TMPRSS2 ;ERGorETV3) 6)t(DDX5;ETV4	Genetic Biomarker Prostate specific antigen Hsp27,Hsp7 0 Cancer stem cells(CTCs)	D D&P D&P D,P& T	FISH Serum/imm unoassay Serum/ELIS A Tumor sample/Imm unocytometr y
Head and neck Squamous cell carcinoma	p57^{KIP2}(Imprinted genes)	Genetic Biomarker	D	Fresh frozen biopsy sample or lymph node or FNAC/PCR
Bladder cancer	CA-19-9 gene	Cancer antigen9- 9(CA19-9) Hsp27,Hsp7 0	D&P D&P	Urine/ELIS A Serum/ELIS A
Osteosarcoma	LOH-LSAMP	Genetic Biomarker HSP Hsp27,Hsp7 0	D D&P	FISH/CGH Serum/ELIS A
Uterine	Heat shock protein genes	Hsp27,Hsp7 0	D&P	Seum/ELIS A
Cervical	Heat shock protein genes	Hsp27,Hsp7 0	D&P	Serum/ELIS A
Squamous cell carcinoma of stomach	APC gene	APC gene	D&P	Blood/tumo rsample/RF LP of chromosom e5q21- 22,Methylat ion status of

					APC gene
Lung Cancer	1)RASSF1(Signal transduction) 2)PR,RAR-β2(Hormonal receptor)-CR5 3)P73,Peg1/MEST(imprinte d genes) Ras- 4)P27(cell cycle) 5)APC,LEF1,TCF4,βcatenin (Wnt/TCF)				
Small cell lung carcinoma NSCLC(non small cell lung carcinoma)	amp(MET gene EML-ALK translocation,EGFR mutation	Genetic Biomarkers	D	Pleural fluid or biopsy/FIS H	
Squamous cell carcinoma oral	amp(EGFR)	Genetic Biomarker	D	Biopsy/FIS H	
Glioblastoma	Retinoblastoma gene amp(EGFR,PDGFRA)	Genetic Biomarker	D	Biopsy/FIS H	

Table 7. Promising Biomarkers in Major Types of Cancer

Type	Sample Type	Detection Method	Biomarkers
Lung Cancer			
LOH	Plasma	PCR	(D21S1245,FHIT)
LOH	Serum	PCR	(ACTBP2,UT762,AR,D3 S4103,D3S1300)
LOH	Bronchial lavage	PCR	(D3S1289,D3S1300,D13 S171,D17S2179E)
LOH, Mutation	Plasma	Plaque hybridization assay for p53 PCR for FHIT(fragile histidine triad gene) and 3p	(p53,2LOH markers)

DNA Methylation	Serum	MSP	(p16,DAPK1,GSTP1,MGMT,RASSFIA,RARB)
DNA methylation	Serum/Plasma	MethyLight	(APC)
DNA Methylation	Plasma	Semi-nested MSP	(p16)
DNA methylation	Serum	F-MSP	(CDKN2A,CDH13)
DNA methylation	Bronchial aspirate	Methylight	(RASSF1A)
RNA marker	Blood	RT-PCR and dot blot	(CEA)
RNA marker	Blood	RTPCR	(preproGRP)
RNA marker	Sputum Blood	RTPCR	(syndecan1,collagen1,ck 19,two novel genes)
RNA marker	Blood	Real-time RTPCR	(LUNX,MUC1,CK19)
RNA marker	Blood	Nested RT-PCR	(BJ-TSA-9,SCC,LUNX)
Protein marker	Serum	Radioimmunoassay	(GRP)
Protein marker	Serum	Radioimmunoassay(CEA,CYFRA 21-1)	(CEA,NSE,CYFRA 21-1)
Breast Cancer			
LOH	Serum	PCR	(D16S421,D10S197,D10 S215,D8S321,D17S849, D14S62,D17S250,D17S8 55,D14S51,D16S421,p5)
LOH	Bone marrow	PCR	(D1S228,D8S321,D10S1 97,D14S51,D14S62,D16 S421,D17S849,D17S85)
LOH, mutation	Plasma	PCR for LOH SSCP for p53	(6LOH markers,p53)
DNA methylation	Plasma	Methylation sensitive restriction digestion and PCR	(p16)
DNA methylation	Serum	MSP	(RASSF1,APC,DAPK1)

DNA methylation	Fine needle aspiration fluid	MSP	(APC,CCND2)
DNA methylation	Plasma	Methylight	(APC,GSTP1,RASSF1A, RARB)
RNA marker	Blood	cDNA array	(SRP19,CD44,TRP-2-86,Maspin,HSIX1,Gro alpha,Myosin light chain,Mdm-2,ZZ38,Beta-tubulin,N33 gene,Lamininα3)
RNA marker	Blood	RT-PCR	(hMAM,βhCG)
RNA Marker	Blood	Real time RT-PCR	(Mammaglobin,B305D,GABAp,B726P)
Type	Sample Type	Detection Method	Biomarkers
Protein marker	Serum	ELISA	(Mammaglobin)
Protein marker	Nipple aspirate fluid	SELDI	(6500Da and 15940Da proteins)
Protein marker	Serum	SELDI	(BC1,BC2,BC3,Proteomic pattern)
Protein marker	Blood	Real time RT-PCR	(Mammaglobin,B305D-C)
Serum protein marker	Serum	Serum Profiling	(RS/DJ1)
Ductal protein marker	Nipple aspirate fluid	Nipple aspirate fluid profiling	(α2-HS-glycoprotein,LipophilinB ,Beta-globin,Hemopexin,Vitamin D –binding Protein)
Autoantibody	Serum/ Plasma	Humoral response	(RS/DJ1,p53,HSP60,HSP 90,Mucin-related)
Ovarian Cancer			
LOH	Serum	PCR	(D5S346,D7S486,D7S522,D11S904,D17S855,D17S579,D17S786,p53)

LOH, Mutation	Peritoneal Wash	PCR for LOH markers Ligation-specific PCR for p53 and K-ras	(D17S579,D17S855,D17 S786,D13S260,D13S267, IL2RB,D22S283,p53,K-Ras)
DNA methylation	Serum	MSP	(BRCA1,RASSFIA,APC, p14,p16,DAPK1)
Protein marker	Serum	Radioimmunoass ay	(OVX1,LASA,CA15-3,CA72-4,CA1215,M-CSF)
Protein marker	Serum	ELISA for Prostasin,Hp-α,SMR Radioimmunoass ay for CA125	(Prostasin,Hp-α,SMR,)
Protein marker	Serum	SELDI	(Apolipoprotein,A1,trunc ated transthyretin,cleaved fragment of inter-α-trypsin inhibitor heavy chain H4)
Protein marker	Serum	ELISA	(Leptin,prolactin,osteopo ntin,insulin-like growth factor II)
Protein marker	Serum	Luminex assay	(IL-6,IL-8,VEGF,EGF,)
Colon Cancer			
DNA marker	Stool	PCR	(K-ras,p53,APC,BAT-26,Long DNA,D2S123,D5S346,D 17S250,BAT25)
DNA marker	Stool	Direct sequence analysis	(APC,K-ras,p53,BAT-26)
DNA methylation	Stool	Modified MSP	(MGMT,CDKN2A, MLH1)
DNA methylation	Stool	MSP	(Vimentin)
RNA marker	Stool	RT-PCR	(c-myc p64,c-myc p67)
Prostate Cancer			
DNA methylation	Plasma/ serum	MethyLight	(GSTP1)

DNA methylation	Plasma	MSP	(GSTP1)
RNA marker	Urine	Real time PCR	(Telomerase)
Protein marker	Serum	SELDI	(4475,5074,5382,7024,78 20,8141,9149,9507,9656 Da Proteins,proteomic pattern)
Protein marker	Plasma	ELISA	(EPCA)
Antibody marker	Serum	ELISA	(AMACR)
Antibody marker	Serum	Protein array/ELISA	(15F1,21H4,7F8,16D12, 24E1,6E2,20F6,8A6,4B9 ,21D10,15H9,24G4,12B6 ,22B1,1B4,21B4,2B10,8 E10,17F10,8D1,18D2,3C 4)
Protein Marker	Serum	Protein array/Spectropho tometry	(KLK2)
Antibody marker	Serum	Protein array/ELISA	(PSMA)
Protein Marker	Serum	Protein array/Spectropho tometry	(KLK11)
Protein marker	Urine	Spectropho-tometry	(PCA3)
Protein Marker	Serum	Protein array	(uPA/uPAR)
DNA marker	Biopsy	CGH	(IGF/IGFBP)
Protein marker	Serum	Protein array	(TGF-β1)
RNA marker	Serum	Protein array	(EZH2)
Protein marker	Serum	Protein array	(PSP94)
Protein marker	Serum	Protein array	(CRISP3)
Protein marker	Serum	Protein array	(Chromogranin A)
Protein marker	Serum	Protein array	(Progastrin releasing peptide)
Protein marker	Serum	Protein array/spectropho tometry	(E-Cadherin)
Protein marker	Serum	Protein array	(Annexin A3)

Protein marker	Serum	Protein array	(PSCA)
Protein marker	Serum	Protein array	(Hepsin)
Protein marker	Serum	Protein array	(IL-6)
Bladder Cancer			
LOH	Urine	PCR	(D4S243,D4S174,FGA, ACTBP2,D8S136,D8S2 58,D8S307,IFNA,D9S12 ,D9S63,D9S162,D9S171 ,D9S283,D9S747,D11S4 88,D11S554THO,D13S8 02,MJD,MJD58,D14S28 8,D14S75,D14S292,D14 S267,D14S258,D16S310 ,D16S476,D17S695,D17 S654,D18S51,D18S364, MBP,D20S48,D21S1245 ,NEFL,GSN,UT762,AB L1,IFNa,)
LOH/MSI	Urine	PCR	(D9S747,D9S171,D9S16 2,IFNA,D4S243)
MSI	Urine/blood	PCR	(Nt2912,D4S243,D9S74 7,D9S171,D17S695,D17 S654)
MSI	Serum	Fluorecent MSA	(D5S1720,D5S476,D8S2 61,D8S560,D9S171,D9S 925,D9S15,D13S153,D1 4S750,D14S61,D14S267 ,D17S799,D17S1306,D2 0S486,D20S607,D20S48 1,D20S480)
MSI	Urine	PCR	(FGA,D4S243,ACTBP2, D9S162,IFN-α,D9S171,D9S747,MJD, D16S310,D16S476,D18 S51,MBP,D21S1245)
DNA Methylation	Urine	MSP	(RARB,DAPK1,Ecad,p1 4,p16,p15,GSTP1,MGM T,RASSFIA,APC)
DNA methylation	Urine	MethyLight	(DAPK1,BCL2,TERT,R ASSF1A,CDH1,APC)

RNA marker	Blood	Nested PCR	(CK20)
RNA marker	Blood	RT-PCR	(CK20,EGFR)
Protein	Urine	Quantitative western blot	(CRT,SNCG,s-COMT)

Genomic Biomarkers

Detection of single nucleotide polymorphisms by PCR, RT-PCR, Nested PCR have today become robust and routine diagnostic for many cancers (Table 5 & 6). They are not only used for diagnosis, but also for disease staging, prognostication but also to decide the genetic predisposition , drug metabolism. More than 30,000- 50,000 SNPs have been identified and their association with different malignant conditions reported and is discussed elsewhere in this chapter. It is pertinent to mention here only ones that are routinely used in the clinic which are:

Gene assays by PCR/RTPCR for:

- JAK2 and
- pharmacogenomics of conventional drugs used in cancer therapy such as
    - Gefitinib for EGFR
    - Cetuxima for K-ras,
    - Irinotecan for UGT1A1,
    - 5FU/Capcitabine for DPD,TS,MTHFR
    - Oxaliplatin /cisplatin/carboplatin for ERCC1
    - Paclitaxeel/docetaxel for ABCB1
    - Pemetrexed for ERCC1,TS,RFC1
    - 6 mercaptapurine for TPMT.

Other tests that are well validated and easy to adapt to present scenario of laboratory medicine are listed below in Table 5 & 6.

Conclusion

The consensus that an effective predictive and safe biomarker is an unmet medical need in cancer diagnostics is, only a goal defined, necessitating to work backward to integrate the identified component parts in the process, plan a path based on the outcome of research. The outcome of research is the discovery of key modulators in the process of the pathobiology of malignancy. The targets of this outcome needs to be translated into measurable readout of clinical pertinence. In the scenario of knowledge being dynamic, this is, neither linear nor simple.

Discovery and clinical application of new biomarkers, is expected to play a significant role in reshaping life science research and life science industry, thereby profoundly influencing the detection and treatment of many diseases and cancer in particular. Clinical oncology is poised to enter a new era in which cancer detection, diagnosis, and treatment will be guided increasingly by the molecular attributes of the individual patient, acquired from several different sources viz. , tumor tissue, host cells/tissues that influence tumor behaviour and body fluids. The resultant panel of biomarkers will not only help the detection and diagnosis, but also answer fundamental questions about biologic behaviour of tumors, resistance to therapy and sensitivity to therapy facilitating individualization of therapy, besides identifying individuals predisposed to cancer. The future of cancer therapy lie in the use of biomarkers that offer the potential to identify and treat cancer years before it is either visible or symptomatic. Exploring the presence of such markers that does not require the tumor tissue to detect them, but are secreted by cancer cells into the blood stream will not only facilitate easy detection without even minimal surgical procedure, but will also be candidates for population based screening.

A large number of biomarkers have been identified by different sequencing technologies in genomics, epigenetic markers, mi RNA, expression profiling, proteomics etc. Most of this data coming from genomics, proteomics and metabolomic profiling will provide a more global view of disease pathogenesis pathway, drug tissue interactions. These surrogate endpoints is likely to add to or supersede existing endpoints . These need to be validated in prospective cohort studies. Translation of these microarray and other DNA based technologies, to clinical practise will require the following issues to be addressed:

- Type of analyte, if it is serum it is fairly straight forward, however if it is tissue biopsy-a need for standardised procedure to obtain biopsy samples is a need
- Standardised protocols for sample transport, storage, and retrieval, irrespective of the testing platform.
- Analytical issues like: type of controls, turnaround time, throughput, detection limit, quantification, on site availability.
- Demonstration of sensitivity, specificity and predictability through clinical validation study.
- An easy, cost effective, accurate and reliable test.
- Analytical performance
- Portability of testing platform
- Cost of replicate analysis, and normalization.
- Training of laboratory technicians.
- Education of ordering Health professional
- Development of clinical and regulatory guidelines
- Last but not least issues of privacy & confidentiality

Thus in the translation of this rapidly expanding field of genomics, the above issues present a barrier to its introduction in clinical practice. The potential for genomic biomarkers to improve diagnosis, prognosis and treatment decisions and long term outcomes can be substantial and provide an insight in disease aetiology that is unprecedented. It is important therefore that all stakeholders involved, work in unison towards the ultimate goal of bringing these biomarkers to clinical practice and add value to patient care. Thus the critical question is who will lead the way to address: road map and critical path initiative.

Abbreviations

APC –Adenomatous Polyposis coli
ALK-Anaplastic lymphoma receptor tyrosine kinase
AMACR-alpha Methylacyl-CoA Racemase
ACTBP2-actin, beta pseudogene 2
BRAF- v-raf murine sarcoma viral oncogene homolog B1
BMPR1A-Bone Morphogenetic Protein receptor type 1A

BAT-26-microsatellite instability marker, a 26-repeat adenine tract located within the fifth intron of the MSH2 gene.

BJ-TSA-9-Lung cancer specific gene, family with sequence similarity 83, member A

CAN-Nucleoporin 214kDa

CREBBP(CBP)- cAMP-regulatory element-binding protein

CBFA2T1(ETO)- Core-binding factor, runt domain, alpha subunit 2; translocated to, 2

CDKN2A-cyclin dependent kinase inhibitor 2A

CDK4-Cyclin dependent kinase4

CCND1-Cyclin D1

COL1A1-Collagen type1,alpha1

CDH-Epithelial cadherin

CEA-Carcinoembryonic antigen

CYFRA 21-1-Cytokeratin 19 fragments

CCND2-Cyclin D2

CD44-Cell adhesion molecule 44

CA-Carcinoma antigen

CRISP3-Cysteine-rich secretory protein

CRT-calreticulin

CK19-Cytokeratin 19

DEK-Oncogene

DAPK-death associated protein kinase

DDX5-dead box helicase 5

DAPK1-Death associated protein kinase1

EVlI- Ena/VASP-like

ETV6-etsvariant6

EGFR-Epidermal growth factor receptor

EGF-Epidermal growth factor

EWSR1- Ewing sarcoma breakpoint region 1.

ERBB2-V-erb-b2 erythroblastic leukemia viral oncogenehomolog2

ER-Estrogen receptor

ESR1-Estrogen receptor 1

EML4-Echinoderm microtubule associated protein like 4

ERG- v-ets erythroblastosis virus E-26 oncogene homolog

EPCA-Early prostate cancer antigen

EZH2-Enhancer of ZesteHomolog2

FOXO1A- Forkhead box O1

FLI1- Friend leukemia integration 1 transcription factor

FHIT-Fragile histidine triad gene
FGA-Fibrinogen alpha
GSTP1-Glutathione S transferase pi
GABAπ-γ-aminobutyrate type A receptor
hCHK2-Checkpoint kinase 2
Her2-Human Epidermal Growth Factor Receptor2
IGH- Immunoglobulin Heavy
IHC-Immunohistochemistry
IL-Interleukin
IGF/IGFBP-Insulin like growth factors,IGF binding proteins
KRAS- v-Ki-ras2Kirsten rat sarcoma viral oncogene homolog
K-ras- human oncogene
KLK2-Kallikrein related peptidase2
LEF1- Lymphoid enhancer-binding factor1
LSAMP- Limbic system associated membrane protein
LUNX- palate, lung and nasal epithelium carcinoma associated
MLL-Mixed Lineage Leukemia
MYST3(MOZ)- MYST Histone acetyltransferase (monocytic leukemia 3)
MLLT3(AF9)- Myeloid/lymphoid or mixed-lineage leukemia translocated
        to, 3
MYH11;CBFB2- Myosin, heavy chain 11, smooth muscle; Core-binding
        factor, beta subunit.
MYC- V-myc myelocytomatosis viral oncogene homolog (avian)
MLLT2- myeloid/lymphoid leukemia translocated to, 2
MYCN- V -myc myelocytomatosis viral related oncogene, neuroblastoma
        derived (avian)
MMAC1-Mutated in Multiple Advanced Cancer 1
MEN1-Multiple endocrine neoplasm type1
MLH1- mutL homolog1
MSH- DNA mismatch repair protein
MET- met proto-oncogene
MGM- Mitochondrial genome maintainance
MBD4- methylated DNA binding domain containing protein
MEST- Mesoderm specific transcript
MDM2- oncogene (mouse double minute 2 homolog)
MSP- Methylation specific PCR
MGMT-O-6 - methylguanine-DNA methyltransferase
MUC1- Mucin1
MLH1- MutL-homolog1

MSI- Microsatellite instability
NPM;RARA- nucleophosmin ; retinoic acid receptor
NF1- Neurofibromatosis
NSE-Neuron specific enolase
PAX3-Paired box3
PTEN- protein tyrosine phosphatise
PIK3CA- phosphatidylinositol-4,5-bisphosphate 3-kinase, catalytic subunit alpha polypeptide
PRKAR1A- Protein kinase, cAMP-dependent, regulatory, type I, alpha
PR- Progesterone receptor
PDGFB- Platelet derived growth factor beta polypeptide
PMS1- Post meiotic segregation Increased 1
PTC- Patched
PDGFRA- platelet derived growth factor receptor, alpha polypeptide
PSMA- Prostate specific membrane antigen
PCA3- Prostate cancer antigen 3
PSP94- Prostate secretory protein
PSCA- Prostate stem cell antigen
ROPN1- rhophilin associated tail protein1
RUNX1- Runt-related transcription factor 1
RET- Rearranged during transfection
RASSF1- Ras association domain containing protein 1
RASSF1A- Ras associated domain family 1
RARB- Retinoic acid receptor beta
SMAD4- SMAD family member4
STK11- Serine/threonine kinase11
SSX- Synovial sarcoma, X breakpoint
SDHD- Succinate dehydrogenase complex, subunit D
SDHC- Succinate dehydrogenase complex, subunit C
SDHB- Succinate dehydrogenase complex, subunit B
SCC- squamous cell carcinoma antigen
SSCP- Single strand conformation analysis
SRP19- Signal recognition particle19kDa protein
SNCG- Synuclein gamma
S-COMT- catechol–o-methyltransferase
TEL - translocation ets leukemia
TCR- T-cell receptor
TP53- Tumor protein 53
TGF-Transforming growth factor

TEP1- TGFb regulated and Epithelial cell enriched Phosphatase 1
TSC1-Tuberous sclerosis
TOP2A- Topoisomerase (DNA) II Alpha
TIMP3- Tissue inhibitor of metalloproteinase
TCF4- Transcription factor 4
TMPRSS2-Transmembrane protease serine2
TGF-β1-Transforming growth factor
TERT-Telomerase reverse transcriptase
uPA/Upar- Urokinase Plasminogen activator,uPAR receptor
VHL- Von Hippel Lindau
VEGF- Vascular endothelial growth factor

References

[1]   Ferlay J, Bray F, Pisani P et al. GLOBOCAN 2000: Cancer incidence, mortality and prevalence worldwide, version 1. 0. IARC Cancer Base No. 5, IARCPress, Lyon, France (2001).

[2]   Cho WSC. Contribution of oncoproteomics to cancer biomarker discovery. *Mol. Cancer.* 2007;6:25.

[3]   Fritz A, Jack A, Parkin DM, Percy C, Shanmugarathan S, Sobin L, Whelan S. International Classification of diseases for oncology (ICD-O). 3rd ed. Geneva: WHO; 2000.

[4]   Robins and Cotran. Neoplasia. Kumar V, Abbas AK, Fausto N, editors. Pathologic Basis of Diseases. Pennsylvania: Elsevier; 2007, 7th ed: p 269-342.

[5]   Caspersson T, Zech L, Johansoon C. Analysis of human metaphase chromosomes by aid of DNA binding fluorescent agents. *Exp. Cell Res.* 1970;49:219–22.

[6]   Drets ME, Shaw MW. Specific banding patterns of human chromosome. *Proc. Natl. Acad. Sci. USA.* 1971;68:2073–77.

[7]   Seabright M. A rapid banding technique for human chromosomes. *Lancet.* 1971;2:971–2.

[8]   Sumner AT, Evans HJ, Buckland RA. New techniques for distinguishing between human chromosomes. *Nature New Biol.* 1971;232:31–2.

[9]   Arrighi FE, Hsu TC. Localization of heterochromatin in human chromosomes. *Cytogenetics.* 1971;10:81–6.

[10] Howell WM, Denton TE, Diamand JR. Differential staining of the satellite regions of human adocentric chromosomes. *Experientia.* 1975;31:260–2.

[11] Casas S, Aventn A, Fuentes F, et al. Genetic diagnosis by genomic hybridization in adult de novo acute myelocytic leukemia. *Cancer Genet Cytogenet.* 2004;153:16–25.

[12] Jarosˇova´ M, Holzerova´ M, Jedlicˇkova´ K, et al. Importance of using comparative genomic hybridization to improve detection of chromosomal changes in childhood acute leukemia. *Cancer Genet Cytogenet.* 2000;123:114–22.

[13] Verdorfer I, Brecevic L, Saul W, et al. Comparative genomic hybridization-aided unraveling of complex karyotypes in human haematopoietic neoplasias. Cancer Genet *Cytogenet.* 2001;124:1–6.

[14] Gall JG. and Pardue ML. *Proc. Natl. Acad. Sci. USA.* 1969;63:378.

[15] Landegent JE, Jansen N, in de Wal, Dirks RW, Baas F and van der Ploeg M, *Hum. Genet.* 1987;77:366.

[16] Lichter P, Cremer T, Borden J, Manuelidis L, and Ward DC, *Hum. Genet.* 1988;80:224.

[17] Pinkel D, Landegent J, Collins C, Fuscoe J, Segraves R, Lucas J, Gray Jw. *Proc. Natl. Acad. Sci. USA.* 1988;85:9138.

[18] Speicher MR, Carter NP. The new cytogenetics: blurring the boundaries with molecular biology. *Nat. Rev. Genet.* 2005;6:782–92.

[19] de Jong H. Visualizing DNA domains and sequences by microscopy: a fifty-year history of molecular cytogenetics. *Genome.* 2003;46:943–6.

[20] Futreal PA, Coin L, Marshall M, Down T, et al. A census of human cancer genes. *Nat. Rev. Cancer.* 2004;4, (3):177-183.

[21] Belaud-Rotureau MA, Parrens M, Dubus P, Garroste JC, de Mascarel A, Merlio JP. A comparative analysis of FISH, RT-PCR, PCR, and immunohistochemistry for the diagnosis of mantle cell lymphomas. *Mod. Pathol.* 2002;15, (5):517-25.

[22] Pinkel D, Straume T, and Gray JW. Cytogenetic analysis using quantitative, high-sensitivity, fluorescence hybridization. *Proc. Natl. Acad. Sci. USA.* 1986;83:2934–38.

[23] Mullink H, Walboomers JM, Tadema TM, Jansen DJ, Meijer CJ. Combined immuno- and non-radioactive hybridocytochemistry on cells and tissue sections: influence of fixation, enzyme pre-treatment, and choice of chromogen on detection of antigen and DNA sequences. *J. Histochem. Cytochem.* 1989;37:603–9.

[24] Speicher MR, Ballard SG, Ward DE. Karyotyping human chromosomes by combinatorial multi-fluor FISH. *Nature Genet.* 1996;12:368–75.

[25] Schrock E, duManoir S, Veldman T, Schoell B, Wienberg J, Ferguson-Smith MA, Ning Y, Ledbetter DH, Bar-Am I, Soenksen D, Garini Y, Ried T. Multicolor spectral karyotyping of human chromosomes. *Science.* 1996;273:494–7.

[26] Tanke HJ, Wiegant J, van Gijlswijk RP, Bezrookove V, Pat-tenier H, Heetebrij RJ, et al. New strategy for multicolor fluorescence in situ hybridisation: COBRA: Combined Binary RAtio labelling. *Eur. J. Hum. Genet.* 1999;7:2–11.

[27] Henderson LJ, Okamoto I, Lestou VS, Ludkovski O, Robichaud M, Chhanabhai M, et al. Delineation of a minimal region of deletion at 6q16. 3 in follicular lymphoma and construction of a bacterial artificial chromosome contig spanning a 6-megabase region of 6q16-q21. *Genes Chromosomes Cancer.* 2004;40:60–65.

[28] Karenko L, Hahtola S, Paivinen S, Karhu R, Syrja S, Kahkonen M, et al. Primary cutaneous T-cell lymphomas show a deletion or translocation affecting NAV3, the human UNC-53 homologue. *Cancer Res.* 2005;65:8101–10.

[29] Elnenaei MO, Hamoudi RA, Swansbury J, Gruszka-Westwood AM, Brito-Babapulle V, Matutes E, et al. Delineation of the minimal region of loss at 13q14 in multiple myeloma. *Genes Chromosomes Cancer.* 2003;36:99–106.

[30] La Starza R, Crescenzi B, Pierini V, Romoli S, Gorello P, Brandimarte L, et al. A common 93-kb duplicated DNA sequence at 1q21. 2 in acute lymphoblastic leukemia and Burkitt lymphoma. *Cancer Genet. Cytogenet.* 2007;175:73–76.

[31] Mancini M, Cedrone M, Diverio D, Emanuel B, Stul M, Vranckx H, et al. Use of dual-color interphase FISH for the detection of inv(16) in acute myeloid leukemia at diagnosis, relapse and during follow-up: a study of 23 patients. *Leukemia.* 2000;14:364–8.

[32] Fejzo MS, Godfrey T, Chen C, Waldman F, and Gray JW. Molecular cytogenetic analysis of consistent abnormalities at 8q12-q22 in breast cancer. *Genes Chromosomes Cancer.* 1998;22:105–13.

[33] Kearney L. Molecular cytogenetics. *Best Pract. Res. Clin. Haematol.* 2001;4:645–69.

[34] Gray JW, Kallioniemi A, Kallioniemi O, Pallavicini M, Waldman F, Pinkel D. Molecular cytogenetics:diagnosis and prognostic assessment. *Curr. Opin. Biotechnol.* 1992;3:623–31.

[35] McNeil N, Ried T. Novel molecular cytogenetic techniques for identifying complex chromosomal rearrangements: technology and applications in molecular medicine. *Expert Rev. Mol. Med.* 2000:2:1–14.

[36] Xu J, Wang N. Identification of chromosomal structural alterations in human ovarian carcinoma cells using combined GTG banding and repetitive fluorescence in situ hybridization (FISH). *Cancer Genet. Cytogenet.* 1994;74:1–7.

[37] Xu J, Tyan T, Cedrone E, Savaraj N, Wang N. Detection of 11q13 amplification as the origin of a homogeneously staining region in small cell lung cancer by chromosome microdissection. *Genes Chromosom Cancer.* 1996;17:172–8.

[38] Dalton SR, Gerami P, Kolaitis NA, Charzan S, Werling R, LeBoit PE, Bastian BC. Use of fluorescence in situ hybridization (FISH) to distinguish intranodal nevus from metastatic melanoma. *Am. J. Surg. Pathol.* 2010;34 (2):231-37.

[39] Gerami P, Mafee M, Lurtsbarapa T, Guitart J, Haghighat Z, Newman M. Sensitivity of fluorescence in situ hybridization for melanoma diagnosis using RREB1, MYB, Cep6, and 11q13 probes in melanoma subtypes. *Arch. Dermatol.* 2010;146, (3):273-78.

[40] Gerami P, Pouryazdanparast P, Vemula S, Bastian BC. Molecular analysis of a case of nevus of ota showing progressive evolution to melanoma with intermediate stages resembling cellular blue nevus. *Am. J. Dermatopathol.* 2010;32, (3):301-5.

[41] Isaac AK, Lertsburapa T, Pathria Mundi J, Martini M, Guitart J, Gerami P. Polyploidy in spitz nevi: a not uncommon karyotypic abnormality identifiable by fluorescence in situ hybridization. *Am. J. Dermatopathol.* 2010;32, (2):144-8.

[42] Halling KC, King W, Sokolova IA, Meyer RG, Burkhardt HM, Halling AC, et al. A comparison of cytology and fluorescence in situ hybridization for the detection of urothelial carcinoma. *J. Uro.* 2000;164, (5):1768-75.

[43] Sokolova IA, Halling KC, Jenkins RB, Burkhardt HM, Meyer RG, Seelig SA, et al. The development of a multitarget, multicolor fluorescence in situ hybridization assay for the detection of urothelial carcinoma in urine. *J. Mol. Diagn.* 2000;2, (3):116-23.

[44] Sholl LM, John Iafrate A, Chou YP, Wu MT, Goan YG, Su L, et al. Validation of chromogenic in situ hybridization for detection of EGFR copy number amplification in non small cell lung carcinoma. *Mod. Pathol.* 2007;20, (10):1028-35.

[45] Simone G, Mangia A, Malfettone A, Rubini V, Siciliano M, Di Benedetto A, et al. Chromogenic in situ hybridization to detect EGFR gene copy number in cell blocks from fine needle aspirates of non small cell lung carcinomas and lung metastases from colo-rectal cancer. *J. Exp. Clin. Cancer Res.* 2010;29:125.

[46] Vorsanova SG, Iurov IuB, Solov'ev IV, Demidova IA, Sharonin VO, Male R, et al. Current methods of molecular cytogenetics in pre- and postnatal diagnosis of chromosome aberrations. *Klin. Lab. Diagn.* 2000;8:36–39.

[47] Divane A, Carter NP, Spathas DH, Ferguson-Smith MA. Rapid prenatal diagnosis of aneuploidy from uncultured amniotic fluid cells using 5-color fluorescence in situ hybridization. *Prenat. Diagn.* 1994;14:1061–69.

[48] Feldman B, Ebrahmin SAD, Hazan SL, Gyi K, Johnson MP, Johnson A, et al. Routine prenatal diagnosis of aneuploidies by FISH studies in high-risk pregnancies. *Am. J. Med. Genet.* 2000;90:233–8.

[49] Heng HH, Squire J, Tsui LC. High resolution mapping of mammalian genes by in situ hybridization to free chromatin. *Proc. Natl. Acad. Sci. USA.* 1992;89:9509–13.

[50] Wiegant J, Kalle W, Mullenders L, Brookes S, Hoovers JMN, Dauwerse JG, et al. High-resolution in situ hybridization using DNA halo preparations. *Hum Mol Genet.* 1992;1:587–592.

[51] Heiskanen M, Peltonen L, Palotie A. Visual mapping by high resolution FISH. *Trends Genet.* 1996;12:379–382.

[52] Raap AK, Florijn RJ, Blonden LAJ, Wiegant J, Vaandrager J-W, Vrolijk H, et al. Fiber FISH as a DNA mapping tool. *Methods Enzymol.* 1996;9:67–73.

[53] Heng HH, Tsui LC. High resolution free chromatin/DNA fiber FISH. *J. Chromatogr.* 1998;806:219–29.

[54] Heng HH, Chamberlain JW, Shi X-M, Spyropoulos B, Tsui L-C, Moens PB. Regulation of meiotic chromatin loop size by chromosomal position. *Proc. Natl. Acad. Sci. USA.* 1996;93:2795–2800.

[55] Heng HH, Spyropoulos B, Moens PB. FISH technology in chromosome and genome research. *Bioessays.* 1997;19:75–84.

[56] Heng HH, Spyropoulos B, Moens PB. DNA-protein in situ co visualization for chromosome analysis. In: Darby IA, editor. Methods in molecular biology: in situ hybridization protocols. Totowa, NJ: Humana Press. 1999.

[57] Weier HU, Greulich-Bode KM, Ito Y, Lersch RA, Fung J. FISH in cancer diagnosis and prognostication: from cause to course of disease. *Expert Rev. Mol. Diagn.* 2002;2, (2):109-19.

[58] Cremer M, Muller S, Kohler D, Brero A, Solovei I. Cell Preparation and Multicolor FISH in 3D Preserved Cultured Mammalian Cells. CSH Protoc. 2007;pdb prot 4723.

[59] Choo KHA. The centromere. Oxford University Press. Oxford. 1997.

[60] Eils R, Uhrig S, Saracoglu K, Satzler K, Bolzer A, Peterson I, Chassery J-M, Ganser M, Speicher MR. An optimized, fully automated system for fast and accurate identification of chromosomal rearrangements by multiplex-FISH (M-FISH). *Cytogenet. Cell Genet.* 1998;82: 160–71.

[61] Castleman KR, Eils R, Morrison L, Piper J, Saracoglu K, Schulze MA, Speicher MR. Classification accuracy in multiple color fluorescence imaging microscopy. *Cytometry.* 2000;4:139–47.

[62] Fan Y-S, Siu V, Jung JH, Xu J. Sensitivity of multiple color spectral karyotyping in detecting small inter-chromosomal rearrangements. *Genet. Testing.* 2000;4:9–14.

[63] Liehr T, Claussen U. Multicolor-FISH approaches for the characterization of human chromosomes in clinical genetics and tumor cytogenetics. *Current Genomics.* 2004;3:213-35.

[64] Lee C, Gisselsson D, Jin C, Nordgren A, Ferguson DO, Blennow E, et al. Limitations of chromosome classification by multicolor karyotyping. *Am. J. Hum. Genet.* 2001;68:1043–47.

[65] Guan X-Y, Meltzer PS, Dalton WS, Trent JM. Identification of cryptic sites of DNA sequence amplification in human breast cancer by chromosome microdissection. *Nat. Genet.* 1994;8:155–61.

[66] Ludecke HJ, Senger G, Claussen U, Horsthemke B. Cloning of defined regions of the human genome by microdissection of banded chromosomes and enzymatic amplification. *Nature.* 1989;338:348–50.

[67] Bohlander SK, Espinosa R III, LeBeau MM, Rowley JD, Diaz MO. A method for the rapid sequence-independent amplification of microdissected chromosomal material. *Genomics.* 1992;13:1322–24.

[68] Carter NP, Ferguson-Smith MA, Perryman MT, Telenius H, Pelmear AH, Leversha MA, et al. Reverse chromosome painting: a method for the rapid analysis of aberrant chromosomes in clinical cytogenetics. *J. Med. Genet.* 1992;29:299–307.

[69] Meltzer PS, Guan X-Y, Burgess A, Trent JM. Rapid generation of region specific probes by chromosome microdissection and their application. *Nature Genet.* 1992;1:24– 28.

[70] Ruano G, Pagliaro EM, Schwartz TP, Lamy K, Messina D, Gaensslen RE, Lee HC. Heat soaked PCR: an efficient method for DNA amplification with applications to forensic analysis. *Bio Tech.* 1992;13:266–74.

[71] Telenius H, Carter NP, Bebb CE, Nordenskjold M, Ponder BAJ, Tunnacliffe A. Degenerate oligonucleotide-primed PCR: general amplification of target DNA by a single degenerate primer. *Genomics.* 1992;13:718–25.

[72] Guan X-Y, Trent JM, Meltzer PS. Generation of band specific painting probes from a single microdissection chromosome. *Hum. Mol. Genet.* 1993;2:1117–21.

[73] Zhang M, Meltzer P, Jenkins R, Guan X-Y, Trent JM. Application of chromosome microdissection probes for elucidation of BCR-ABL fusion and variant Philadelphia chromosome translocations in chronic myelogenous leukemia. *Blood.* 1993;81:3365– 71.

[74] Xu J, Cedrone E, Roberts M, Wu G, Gershagen S, Wang N. The characterization of chromosomal rearrangements by a combined micro-FISH approach in a patient with myelodysplastic syndrome. *Cancer Genet. Cytogenet.* 1995;83:105–10.

[75] Xu J, Fong C-T, Cedrone E, Sullivan J, Wang N. Prenatal identification of denovo marker chromosomes using micro- FISH approach. *Clin. Genet.* 1998;53:490– 96.

[76] Abeysinghe H, Cedrone E, Tyan T, Xu J, Wang N. Amplification of C-MYC as the origin of the homogeneous staining region in ovarian carcinoma detected by micro- FISH. *Cancer Genet. Cytogenet.* 1999;114:136– 43.

[77] Sen S, Sen P, Mulac-Jericevic B, Zhou H, Pirrotta V, Stass S. Microdissected double minute DNA detects variable patterns of chromosomal localization and multiple abundantly expressed transcripts in normal and leukemic cells. *Genomics.* 1994;19:542–51.

[78] Tanner M, Gancberg D, di Leo A, Larsimont D, Rouas G, Piccart MJ, et al. Chromogenic in situ hybridization: a practical alternative for fluorescence in situ hybridization to detect HeR-2/neu oncogene amplification in archival breast cancer samples. *Am. J. Pathol.* 2000;157(5):1467-72.

[79] Mitelman F, Mertens F, Johansson B. Prevalence estimates of recurrent balanced cytogenetic aberrations and gene fusions in unselected patients with neoplastic disorders. Genes, *Chromosomes Cancer.* 2005;43:350–66.

[80] Jobanputra V, Kripalani A, Chaudhry VP, Kucheria K. Detection of chromosomal abnormalities using fluorescence in situ hybridization (FISH). *Natl. Med. J. India.* 1998;1:259-263.

[81] Domenice S, Nishi MY, Billerbeck AE, Carvalho FM, Frade EM, Latronico AC, Arnhold IJ, Mendonca BB. Molecular analysis of SRY gene in Brazilian 46, XX sex reversed patients: absence of SRY sequence in gonadal tissue. *Med. Sci. Monit.* 2001;7:238–41.

[82] Hochhaus A, Weisser A, La Rosee P, Emig M, Muller MC, Saussele S, Reiter A, Kuhn C, Berger U, Hehlmann R, Cross NC. Detection and quantification of residual disease in chronic myelogenous leukemia. *Leukemia.* 2000;14:998–1005.

[83] Froncillo MCC, Maffei L, Cantonetti M, Poeta GD, Lentini R, Bruno A, et al. FISH analysis for CML monitoring? *Ann. Hematol.* 1996;73:113–19.

[84] Coe BP, Ylstra B, Carvalho B, et al. Resolving the resolution of array CGH. *Genomics.* 2007;89:647–53.

[85] Stoler DL, Chen N, Basik M, et al. The onset and extent of genomic instability in sporadic colorectal tumor progression. *Proc. Natl. Acad. Sci. USA.* 1999;96:15121–26.

[86] Ried T, Knutzen R, Steinbeck R, et al. Comparative genomic hybridization reveals a specific pattern of chromosomal gains and losses during the genesis of colorectal tumors. Genes, *Chromosomes Cancer.* 1996;15:234–45.

[87] Meijer GA, Hermsen MAJA, Baak JPA, et al. Progression from colorectal adenoma to carcinoma is associated with non-random chromosomal gains as detected by CGH. *J. Clin. Pathol.* 1998;51:901–9.

[88] Bardi G, Sukhikh T, Pandis N, Fenger C, Kronborg O, Heim S. Karyotypic characterization of colorectal adenocarcinomas. *Genes Chromosomes Cancer.* 1995;12:97–109.

[89] Shih IM, Zhou W, Goodman SN, Lengauer C, Kinzler KW, Vogelstein B. Evidence that genetic instability occurs at an early stage of colorectal tumorigenesis. *Cancer Res.* 2001;61:818– 22.

[90] Ried T, Heselmeyer-Haddad K, Blegen H, Schrock E, Auer G. Genomic changes defining the genesis, progression, and malignancy potential in solid human tumors, a phenotype/genotype correlation. *Genes Chromosomes Cancer.* 1999;25:195–204.

[91] Kallioniemi A, Kallioniemi OP, Sudar D, et al. Comparative genomic hybridization for molecular cytogenetic analysis of solid tumors. *Science.* 1992;258:818–21.

[92]   Knuutila S, Bjorkqvist A-M, Autio K, Tarkkanen M, Wolf M, Monni O, et al. DNA copy number amplifications in human neoplasms: review of comparative genomic hybridization studies. *Am. J. Pathol.* 1998;152:1107.

[93]   Weiss MM, Hermsen MA, Meijer GA, et al. Comparative genomic hybridization. *Mol. Pathol.* 1999;52:243–51.

[94]   Kirchhoff M, Gerdes T, Rose H, Maahr J, Ottesen AM, Lundsteen C. Detection of chromosomal gains and losses in comparative genomic hybridization analysis based on standard reference intervals. *Cytometry.* 1998;31:163-73.

[95]   Kirchhoff M, Rose H, Lundsteen C. High resolution comparative genomic hybridization in clinical cytogenetics. *F. Med. Genet.* 2001;38:740-44.

[96]   Lichter P, Joos S, Bentz M, Lampel S. Comparative genomic hybridization: uses and limitations. *Semin. Hematol.* 2000;37:348-57.

[97]   Mantripragada KK, Buckley PG, de Sta°hl TD and Dumanski JP. Genomic microarrays in the spotlight. *Trends Genet.* 2004;20(2):87–94.

[98]   Carter NP. Methods and strategies for analyzing copy number variation using DNA microarrays. *Nat. Genet.* 2007;39:S16–S21.

[99]   Kirchoff M, Gerdes T, Maahr J, Rose H, Bentz M, Dohner H, Lundsteen C. Deletions below 10 megabasepairs are detected in comparative genomic hybridization by standard reference intervals. *Genes Chromosomes cancer.* 1999;25:41-413.

[100] Pinkel D, Segraves R, Sudar D, Clark S, Poole I, Kowbel D, et al. High resolution analysis of DNA copy number variation using comparative genomic hybridization to microarrays. *Nat. Genet.* 1998;20:207–11.

[101] Solinas-Toldo S, Lampel S, Stilgenbauer S, Nickolenko J, Benner A, Dohner H, et al. Matrix-based comparative genomic hybridization: biochips to screen for genomic imbalances. *Genes Chromosomes Cancer.* 1997;20:399–407.

[102] Pollack JR, Perou CM, Alizadeh AA, Eisen MB, Pergamenschikov A, Williams CF, et al. Genome-wide analysis of DNA copy-number changes using cDNA microarrays. *Nat. Genet.* 1999;23:41–46.

[103] Snijders AM, Nowak N, Segraves R, et al. Assembly of microarrays for genome-wide measurement of DNA copy number. *Nat. Genet.* 2001;29:263–64.

[104] Shaffer LG, Kennedy GM, Spikes AS, Lupski AS JR. Diagnosis of CMT1A duplications and HNPP deletions by interphase FISH:

implications for testing in the cytogenetics laboratory. *Am. F. Med. Genet.* 1997;69:325-31.

[105] Ballif BC, Rorem EA, Sundin K, Lincicum M, Gaskin S, et al. Detection of low level mosaicism by array CGH in routine diagnostic specimens. *Am. F. Med. Genet.* 2006; A140: 2757-67.

[106] Cheung SW, Shaw CA, Scott DA, Patel A, Sahoo T, et al. Microarray-based CGH detects chromosomal mosaicism not revealed by conventional cytogenetics. *Am. F. Med. Genet.* 2007; A 143: 1679-86.

[107] Saxon PJ, Srivatsan ES, Standbridge EJ. Introduction of human chromosome 11 via microcell transfer controls tumorigenic expression of Hela cells. *EMBO J.* 1986;5:3461–66.

[108] Oshimura M, Kugoh H, Koi M, Shimizu M, Yamada H, Satoh H, Barrett JC. Transfer of a normal human chromosome 11 suppresses tumorigenicity of some but not all tumor cell lines. *J. Cell Biochem.* 1990;42:135–142.

[109] Yamada H, Wake N, Fujimoto S, Barrett JC, Oshimura M. Multiple chromosomes carrying tumor suppressor activity for a uterine endometrial carcinoma cell line identified by microcell-mediated chromosome transfer. *Oncogene.* 1990;5:1141–47.

[110] Poignee M, Backsch C, Beer K, Jansen L, Wagenbach N, Stanbridge EJ, et al. Evidence for a putative senescence gene locus within the chromosomal region 10p14-p15. *Cancer Res.* 2001;61:7118–21.

[111] Sandhu AK, Kaur GP, Reedy DE, Rane NS, Athwal RS. A gene on 6q14-21 restores senescence to immortal ovarian tumor cells. *Oncogene.* 1996;12:247–52.

[112] Weissman BE, Saxon PJ, Pasquale SR, Jones GR, Geiser AG, Stanbridge EJ. Introduction of a normal chromosome 11 into a Wilm's tumor cell line controls its tumorigenic expression. *Science.* 1987;236:175–80.

[113] Negrini M, Castagnoli A, Sabbioni S, Recanatini E, Giovannini G, Possati L, et al. Suppression of tumorigenesis by the breast cancer cell line MCF-7 following transfer of a normal human chromosome 11. *Oncogene.* 1992;7:2013–18.

[114] Negrini M, Sabbioni S, Possati L, Rattan S, Corallini A, Barbanti-Brodano G, Croce CM. Suppression of tumorigenicity of breast cancer cells by microcell-mediated chromosome transfer: studies on chromosome6 and 11. *Cancer.* 1994;54:1331–36.

[115] Rimessi P, Gualandi F, Morelli C, Trabanelli C, Wu Q, Possati L, et al. Transfer of human chromosome 3 to an ovarian carcinoma cell line

identified three regions on 3p involved in ovarian cancer. *Oncogene.* 1994;9:3467–74.

[116] Theile M, Hartmann S, Scherthan H, Arnold W, Deppert W, Frege R, et al. Suppression of tumorigenicity of breast cancer cells by transfer of human chromosome 17 does not require transferred BRCA1 and p53 genes. *Oncogene.* 1995;10:439–47.

[117] Cao Q, Abeysinghe H, Chow O, Xu J, Kaung HL, Fong CF, et al. Suppression of tumorigenicity on human ovarian carcinoma cell line SKOV-3 by microcell mediated transfer of a chromosome 11. *Cancer Genet. Cytogenet.* 2001;129:131–7.

[118] Yoshida BA, Sokoloff MM, Welch DDR, Rinker- Schaeffer CW. Metastasis-suppressor genes: a review and perspective on an emerging field. *J. Natl. Cancer Inst.* 2000;92(21):1717–30.

[119] Cuthbert AP, Bond J, Trott DA, Gill S, Broni J, Marriott A, et al. Telomerase repressor sequences on chromosome 3 and induction of permanent growth arrest in human breast cancer cells. *J. Natl. Cancer Inst.* 1999; 91: 37–45.

[120] Steenbergen RDM, Kramer D, Meijer CJLM, Walboomers JMM, Trott DA, Cuthbert AP, Newbold RF, Overkamp WJL, Zdzienicka MA, Snijders PJF. Telomerase suppression by chromosome 6 in a human papillomavirus type 16–immortalized keratinocyte cell line and in a cervical cancer cell line. *J. Natl. Cancer Inst.* 2001; 93(11): 865–72.

[121] Carvalho B, Ouwerkerk E, Meijer GA, Ylstra B. High resolution microarray comparative genomic hybridisation analysis using spotted oligonucleotides. *J. Clin. Pathol.* 2004; 57(6): 644–6.

[122] De Lellis L, Curia MC, Aceto GM, Toracchio S, Colucci G, Russo A, et al. Analysis of extended genomic rearrangements in oncological research. *Ann. Oncol.* 2007; 18(Suppl 6): vi173– vi178.

[123] Maciejewski JP, Tiu RV, O'Keefe C. Application of array-based whole genome scanning technologies as a cytogenetic tool in hematological malignancies. *Br. J. Haematol.* 2009;146(5), 479–88.

[124] Lucito R, Healy J, Alexander J, et al. : Representational oligonucleotide microarray analysis: a high-resolution method to detect genome copy number variation. *Genome Res.* 2003;13:2291–305.

[125] Lisitsyn N, Wigler M. Cloning the differences between two complex genomes. *Science.* 1993,259:946–51.

[126] Sebat J, Lakshmi B, Troge J, et al. Large-scale copy number polymorphism in human genome. *Science.* 2004;305:525–8.

[127] Gunderson KL, Steemers FJ, Lee G, Mendoza LG and Chee MS. A genome-wide scalable SNP genotyping assay using microarray technology. *Nat. Genet.* 2005; 37: 549–54.

[128] Bignell GR, Huang J, Greshock J, Watt S, Butler A, West S, et al. High-resolution analysis of DNA copy number using oligonucleotide microarrays. *Genome Res.* 2004;14: 287–95.

[129] Heinrichs S, and Look AT. Identification of structural aberrations in cancer by SNP array analysis. *Genome Biol.* 2007; 8: 219.

[130] LaFramboise T, Weir BA, Zhao X, Beroukhim R, Li C, Harrington D, et al. Allele-specific amplification in cancer revealed by SNP array analysis. *PLoS Comput. Biol.* 2005;1: e65.

[131] Park JT, Li M, Nakayama K, Mao TL, Davidson B, Zhang Z, et al. Notch3 gene amplification in ovarian cancer. *Cancer Res.* 2006;66:6312–18.

[132] Gorringe KL, Jacobs S, Thompson ER, Sridhar A, Qiu W, Choong DY, et al. High-resolution single nucleotide polymorphism array analysis of epithelial ovarian cancer reveals numerous microdeletions and amplifications. *Clin. Cancer Res.* 2007;13:4731–9.

[133] Rubin MA, Varambally S, Beroukhim R, Tomlins SA, Rhodes DR, Paris PL, et al. Overexpression, amplification, and androgen regulation of TPD52 in prostate cancer. *Cancer Res.* 2004;64:3814–22.

[134] Koochekpour S, Zhuang YJ, Beroukhim R, Hsieh CL, Hofer MD, Zhau HE, et al. Amplification and over expression of prosaposin in prostate cancer. *Genes Chromosomes Cancer.* 2005;44:351–64.

[135] Tsafrir D, Bacolod M, Selvanayagam Z, Tsafrir I, Shia J, Zeng Z, et al. Relationship of gene expression and chromosomal abnormalities in colorectal cancer. *Cancer Res.* 2006;66:2129–37.

[136] Garraway LA, Widlund HR, Rubin MA, Getz G, Berger AJ, Ramaswamy S, et al. Integrative genomic analyses identify MITF as a lineage survival oncogene amplified in malignant melanoma. *Nature.* 2005;436:117–22.

[137] Harada T, Chelala C, Bhakta V, Chaplin T, Caulee K, Baril P, et al. Genome-wide DNA copy number analysis in pancreatic cancer using high-density single nucleotide polymorphism arrays. *Oncogene.* 2007

[138] Baross A, Delaney AD, Li HI, Nayar T, Flibotte S, Qian H, et al. Assessment of algorithms for high throughput detection of genomic copy number variation in oligonucleotide microarray data. *BMC Bioinformatics.* 2007;8:368.

[139] LaFramboise T. Single nucleotide polymorphism arrays: a decade of biological, computational and technological advances. *Nucleic Acids Res.* 2009;37(13):4181–93.

[140] Tiu RV, Gondek LP, O'Keefe CL et al. Prognostic impact of SNP array karyotyping in myelodysplastic syndromes and related myeloid malignancies. *Blood.* 2011;117(17), 4552-60.

[141] Irizarry RA, Bolstad BM, Collin F, Cope LM, Hobbs B, Speed TP. Summaries of Affymetrix GeneChip probe level data. *Nucleic Acids Res.* 2003;31: E15.

[142] Golub TR, Slonim DK, Tamayo P, et al. Molecular classification of cancer, class discovery and class prediction by gene expression monitoring. *Science.* 1999;286:531–37.

[143] Su YA, Bittner ML, Chen Y, et al. Identification of tumor-suppressor genes using human melanoma cell lines UACC903, UACC903(+6), and SRS3 by comparison of expression profiles. *Mol. Carcinog.* 2000;28:119–27.

[144] Maniotis AJ, Folberg R, Hess A, et al. Vascular channel formation by human melanoma cells in vivo and in vitro, vasculogenic mimicry . *Am. J. Pathol.* 1999;155:739–52.

[145] Brasier AR. Retriever and Compare Table, two informatics tools for data analysis of high-density oligonucleotide arrays. *Biotechniques.* 2002;32:100–109.

[146] Mutter GL, Baak JPA, Fitzgerald J, et al. Global expression changes of constitutive and hormonally regulated genes during endometrial neoplastic transformation. *Gynecol. Oncol.* 2001;83:177–85.

[147] Pruitt KD, Katz KS, Sicotte H, Maglott DR. Introducing RefSeq and LocusLink, curated human genome resources at the NCBI. *Trends Genet.* 2000;16:44–47.

[148] Fiegler H, Gribble SM, Burford DC, Carr P, Prigmore E, Porter KM, et al. Array painting: a method for the rapid analysis of aberrant chromosomes using DNA microarrays. *J. Med. Genet.* 2003;40:664–70.

[149] Wang TL, Maierhofer C, Speicher MR, et al. Digital karyotyping. *Proc. Natl. Acad. Sci. USA.* 2002;99:16156-61.

[150] Velculescu VE, Zhang L, Vogelstein B, et al. Serial analysis of gene expression. *Science.* 1995;270:484–7.

[151] Saha S, Sparks AB, Rago C, et al. Using the transcriptome to annotate the genome. *Nat. Biotechnol.* 2002;20:508–12.

[152] Nakayama K, Nakayama N, Davidson B, Katabuchi H, Kurman RJ, Velculescu VE, et al. Homozygous deletion of MKK4 in ovarian serous carcinoma. *Cancer Biol. Ther.* 2006;5:630– 4.

[153] Shih IM, Sheu JJ, Santillan A, Nakayama K, Yen MJ, Bristow RE, et al. Amplification of a chromatin remodelling gene, Rsf-1/HBXAP, in ovarian carcinoma. *Proc. Natl. Acad. Sci. USA.* 2005;102:14004–9.

[154] Wang TL, Diaz LA, Jr Romans K, Bardelli A, Saha S, Galizia G, et al. Digital karyotyping identifies thymidylate synthase amplification as a mechanism of resistance to 5-fluorouracil in metastatic colorectal cancer patients. *Proc. Natl. Acad. Sci. USA.* 2004;101:3089–94.

[155] Korner H, Epanchintsev A, Berking C, Schuler-Thurner B, Speicher MR, Menssen A, et al. Digital karyotyping reveals frequent inactivation of the dystrophin/ DMD gene in malignant melanoma. *Cell Cycle.* 2007;6:189–98.

[156] Di C, Liao S, Adamson DC, Parrett TJ, Broderick DK, Shi Q, et al. Identification of OTX2 as a medulloblastoma oncogene whose product can be targeted by all-trans retinoic acid. *Cancer Res.* 2005;65:919–24.

[157] Shih IM, and Wang TL. Apply innovative technologies to explore cancer genome. *Curr. Opin. Oncol.* 2005;17:33–38.

[158] Smiraglia DJ, Plass C. The study of aberrant methylation in cancer via restriction landmark genomic scanning. *Oncogene.* 2002;21(5):5414-26.

[159] Rhee I, Bachman KE, Park BH et al. DNMT 1 and DNMT3b co-operate to silence genes in human cancer cells. *Nature.* 2002;416(6880):552-6.

[160] Higuchi R, Dollinger G, Walsh PS, Griffith R. Simultaneous amplification and detection of specific DNA sequences. *Biotechnology* (N Y). 1992;10:413–17.

[161] De Preter K, Speleman F, Combaret V, Lunec J, Laureys G, Eussen BH, et al. Quantification of MYCN, DDX1, and NAG gene copy number in neuroblastoma using a real-time quantitative PCR assay. *Mod. Pathol.* 2002;15:159–66.

[162] Boensch M, Oberthuer A, Fischer M, Skowron M, Oestreich J, Berthold F, Spitz R. Quantitative real-time PCR for quick simultaneous determination of therapy-stratifying markers MYCN amplification, deletion 1p and 11q. *Diagn. Mol. Pathol.* 2005;14:177–182.

[163] Armour JA, Barton DE, Cockburn DJ and Taylor GR. The detection of large deletions or duplications in genomic DNA. *Hum. Mutat.* 2002;20:325–37.

[164] Schouten JP, McElgunn CJ, Waaijer R, Zwijnenburg D, Diepvens F, and Pals G. Relative quantification of 40 nucleic acid sequences by

multiplex ligation-dependent probe amplification. *Nucleic Acids Res.* 2002;30:e57.

[165] Armour JA, Sismani C, Patsalis PC and Cross G. Measurement of copy number by hybridization with amplifiable probes. *Nucleic Acids Res.* 2000;28:605–9.

[166] De Lellis L, Curia MC, Catalano T, De Toffol S, Bassi C, Mareni C, et al. Combined use of MLPA and nonfluorescent multiplex PCR analysis by high performance liquid chromatography for the detection of genomic rearrangements. *Hum. Mutat.* 2006;27:1047–56.

[167] Turner DJ, Shendure J, Porreca G, Church G, Green P, Tyler- Smith C, Hurles ME. Assaying chromosomal inversions by single-molecule haplotyping. *Nat. Methods.* 2006;3:439–45.

[168] Margulies M, et al. Genome sequencing in microfabricated high-density picolitre reactors. *Nature.* 2005;437:376–80.

[169] Tawfik D S, Griffiths AD. Man-made cell-like compartments for molecular evolution. *Nat. Biotechnol.* 1998;16:652–6.

[170] Sanger F, Nicklen S, Coulson A R. DNA sequencing with chain-terminating inhibitors. *Proc. Natl. Acad. Sci. USA.* 1977;74:5463–67.

[171] Smith L M, et al. Fluorescence detection in automated DNA sequence analysis. *Nature.* 1986;321:674–9.

[172] Prober J M, et al. , A system for rapid DNA sequencing with fluorescent chain terminating dideoxynucleotides. *Science.* 1987;238:336–41.

[173] Cohen A S, Najarian D R, Paulus A, Guttman A, Smith J A, Karger B L. Rapid separation and purification of oligonucleotides by high-performance capillary gel electrophoresis. *Proc. Natl. Acad. Sci. USA.* 1988;85:9660–3.

[174] Hall N. Advanced sequencing technologies and their wider impact in microbiology. *J. Exp. Biol.* 2007;210:1518–25.

[175] Stratton M R, Campbell P J & Futreal P A. The cancer genome. *Nature.* 2009;458:719–24.

[176] Wheeler DA, Srinivasan M, Egholm M et al. The complete genome of an individual by massively parallel DNA sequencing. *Nature.* 2008;452 (7189):872–76.

[177] Peiffer DA, Le J M, Steemers FJ et al. High-resolution genomic profiling of chromosomal aberrations using Infinium whole-genome genotyping. *Genome Res.* 2006;16(9):1136–48.

[178] Olshen AB, Venkatraman ES, Lucito R, Wigler M. Circular binary segmentation for the analysis of array-based DNA copy number data. *Biostatistics.* 2004;5(4):557–72.

[179] AnsorgeWJ. Next-generation DNA sequencing techniques. *New Biotechnol.* 2009;25:195–203.

[180] Shendure J, Ji H. Next-generation DNA sequencing. *Nat. Biotechnol.* 2008;26:1135–45.

[181] Thomas RK, Nickerson E, Simons JF, Janne PA, Tengs T, Yuza Y, et al. Sensitive mutation detection in heterogeneous cancer specimens by massively parallel picoliter reactor sequencing. *Nat. Med.* 2006; 12: 852–5.

[182] Pleasance ED, Cheetham RK, Stephens PJ et al. A comprehensive catalogue of somatic mutations from a human cancer genome. *Nature.* 2010;463(7278), 191–6.

[183] Pleasance ED, Stephens PJ, O'Meara S et al. A small-cell lung cancer genome with complex signatures of tobacco exposure. *Nature.* 2010;463(7278), 184–190.

[184] Bennett S. Solexa Ltd, *Pharmacogenomics.* 2004;5:433–8.

[185] Bennett ST, Barnes C, Cox A, Davies L, Brown C. Toward the 1,000 dollars human genome. *Pharmacogenomics.* 2005;6:373–82.

[186] Bentley DR. Whole-genome re-sequencing. *Curr. Opin. Genet. Dev.* 2006;16:545–52.

[187] Bentley DR, Balasubramanian S, SwerdlowHP et al. Accurate whole human genome sequencing using reversible terminator chemistry. *Nature.* 2008;456(7218), 53–59.

[188] Hutchison III CA. DNA sequencing: bench to bedside and beyond. *Nucleic Acids Res.* 2007;35:6227–37.

[189] Mardis ER. Next-generation DNA sequencing methods. *Annu. Rev. Genomics Hum Genet.* 2008;9:387–402.

[190] Metzker ML. Sequencing technologies– the next generation. *Nat. Rev. Genet.* 2010;11(1), 31–46.

[191] Ley T J. et al. DNA sequencing of a cytogenetically normal acute myeloid leukemia genome. *Nature.* 2008;456:66–72.

[192] Pinard, R. et al. Assessment of whole genome amplification-induced bias through high-throughput, massively parallel whole genome sequencing. *BMC Genomics.* 2006;7:216.

[193] Beck C R. et al. LINE-1 retrotransposition activity in human genomes. *Cell.* 2010;141:1159–70.

[194] Huang C R. et al. Mobile interspersed repeats are major structural variants in the human genome. *Cell.* 2010;141:1171–82.

[195] Leary RJ, Kinde I, Diehl F, Schmidt K, Clouser C, Duncan, Antipova A, et al. Development of personalized tumor biomarkers using massively parallel sequencing. *Sci. Transl. Med.* 2010;2:20ra14.

[196] Campbell, P. J. et al. Identification of somatically acquired rearrangements in cancer using genome-wide massively parallel paired-end sequencing. *Nature Genet.* 2008;40:722–729.

[197] Chiang D. Y. et al. High-resolution mapping of copynumber alterations with massively parallel sequencing. *Nature Methods.* 2009;6:99–103.

[198] Huang YW, Huang TH, Wang LS. Profiling DNA methylomes from microarray to genome-scale sequencing. *Technol. Cancer Res. Treat.* 2010;9:139–47.

[199] Lister R, Pelizzola M, Dowen RH, Hawkins RD, Hon G, et al. Human DNA methylomes at base resolution show widespread epigenomic differences. *Nature.* 2009;462:315–22.

[200] Neff T and Armstrong SA. Chromatin maps, histone modifications and leukemia. *Leukemia.* 2009;23:1243–51.

[201] Mortazavi A,Williams BA, McCue K, Schaeffer L, Wold B. Mapping and quantifying mammalian transcriptomes by RNA-seq. *Nat. Methods.* 2008;5:621–28.

[202] Wang Z, Gerstein M, Snyder M. RNA-seq: a revolutionary tool for transcriptomics. *Nat. Rev. Genet.* 2009;10:57–63.

[203] Nagalakshmi U, Wang Z, Waern K, Shou C, Raha D, et al. The transcriptional landscape of the yeast genome defined by RNA sequencing. *Science.* 2008;320:1344–49.

[204] Trapnell C, Pachter L, Salzberg SL. TopHat: discovering splice junctions with RNA-seq. *Bioinformatics.* 2009;25:1105–11.

[205] Maher CA, Kumar-Sinha C, Cao X, Kalyana-Sundaram S, Han B, et al. Transcriptome sequencing to detect gene fusions in cancer. *Nature.* 2009;458:97–101

[206] Morozova O, Hirst M, Marra MA. Applications of new sequencing technologies for transcriptome analysis. *Annu. Rev. Genomics Hum. Genet.* 2009;10:135–51.

[207] Mamanova L, Coffey AJ, Scott CE, Kozarewa I, Turner EH, Kumar A, et al. Target-enrichment strategies for next-generation sequencing. *Nat. Methods.* 2010;7:111-118.

[208] Tuzun, E, Sharp, AJ, Bailey JA, Kaul R, Morrison VA, Pertz LM, et al. Fine-scale structural variation of the human genome. *Nat. Genet.* 2005;37:727–732.

[209] Freeman JL, Perry GH, Feuk L, Redon R, McCarroll SA, Altshuler DM, et al. Copy number variation: new insights in genome diversity. *Genome Res.* 2006;16:949–961.

[210] Volik S, Zhao S, Chin K, Brebner JH, Herndon DR, Tao Q, et al. End-sequence profiling: sequence-based analysis of aberrant genomes. *Proc. Natl. Acad. Sci. USA.* 2003;100:7696–7701.

[211] Volik S, Raphael BJ, Huang G, Stratton MR, Bignel G, Murnane J, et al. Decoding the fine scale structure of a breast cancer genome and transcriptome. *Genome Res.* 2006;16:394–404.

[212] Krzywinski M, Bosdet I, Mathewson C, Wye N, Brebner J, Chiu R, et al. A BAC clone fingerprinting approach to the detection of human genome rearrangements. *Genome Biol.* 2007;8:R224.

[213] Mardis ER. Anticipating the 1,000 dollar genome. *Genome Biol.* 2006;7:112.

[214] Eid J, Fehr A, Gray J, Luong K, Lyle J, Otto G, et al. Real-time DNA sequencing from single polymerase molecules. *Science.* 2009;323:133–8.

[215] Pourmand N, Karhanek M, Persson HH, Webb CD, Lee TH, Zahradnikova A, Davis RW. Direct electrical detection of DNA synthesis. *Proc. Natl. Acad. Sci. U S A.* 2006;103:6466–70.

[216] Reynolds T. For proteomics research, a new race has begun. *J. Natl. Cancer Inst.* 2002, 94, 552–4.

[217] Czerwenka KF, Manavi M, Hosmann J, et al. Comparative analysis of two-dimensional protein patterns in malignant and normal human breast tissue. *Cancer Detect. Prev.* 2001;25: 268–79.

[218] PALUMBO KS, WANDS JR, SAFRAN H et al. : Human aspartyl (asparaginyl) betahydroxylase monoclonal antibodies: potential biomarkers for pancreatic carcinoma. *Pancreas.* 2002; 25:39-44.

[219] LOO C, LOWERY A, HALAS N et al. : Immunotargeted nanoshells for integrated cancer imaging and therapy. *Nano Lett.* 2005;5:709-11.

[220] Ng PC, Henikoff S. Accounting for human polymorphisms predicted to affect protein function. *Genome Res.* 2002;12:436–46.

[221] Ramensky V, Bork P, Sunyaev S. Human non-synonymous SNPs: server and survey. *Nucleic Acids Res.* 2002;30:3894–900.

[222] Sunyaev S, Ramensky V, Koch I, LatheW3rd, Kondrashov AS, Bork P. Prediction of deleterious human alleles. *Hum. Mol. Genet.* 2001;10:591–97.

[223] Wood LD, Parsons DW, Jones S, Lin J, Sjoblom T, et al. The genomic landscapes of human Breast and colorectal cancers. *Science.* 2007;318:1108–13.

[224] Cohen S. Isolation of a Mouse Sub-maxillary Gland Protein Accelerating Incisor Eruption and Eyelid Opening in the New-Born Animal. *J. Biol. chem.* 1962;237:555-62.

[225] Balk S, Whitfield F, Youdale T, Braun AC. Roles of Calcium, Serum, Plasma and Folic Acid in the Control of Proliferation of Normal and Rous Sarcoma Virus Infected Chicken Fibroblasts. *Proc. Nat. Acad. Sci.* 1973;70:675-9.

[226] Ross R, Glomset JA, Kariya B, Harker L. Platelet Dependent Serum Factor that Stimulates the Proliferation of Arterial Smooth Muscle Cells in Vitro. *Proc. Nat. Acad. Sci.* 1974;71:1207-10.

[227] Kohler N and Lipton A. Platelets as a Source of Fibroblast Growth Promoting Activity. *Exp. Cell Res.* 1974;87:297-301.

[228] Pledger WJ, Stiles CD, Antoniades HN, Scher CD. Induction of DNA Synthesis in BALB/c 3T3 Cells by serum components; re evaluation of the commitment process. *Proc. Nat. Acad. Sci.* 1977;74;4481-85.

[229] Pledger WJ, Stiles CD, Antoniades HN, Scher CD. An Ordered Sequence of Events Is Required Before BALB/c-3T3 Cells Become Committed to DNA Synthesis. *Proc. Nat. Acad. Sci.* 1978;75:2839-43.

[230] Antoniades HN,Scher CD,Stiles CD. Purification of Human Platelet-Derived Growth Factor. *Proc. Nat. Acad. Sci.* 1979;76:1809-13.

[231] Gray A, Dull TJ, Ullrich A. Nucleotide sequence of Epidermal Growth Factor cDNA Predicts a 128,000-Molecular weight Protein Precursor. *Nature.* 1983;303:722-5.

[232] Scott J, Urdea M, Quiroga M, Sanchez-Pescador R, Fong N, Selby M, et al. Structure of a Mouse Submaxillary Messanger RNA Encoding Epidermal Growth Factor and Seven Related Proteins. *Science.* 1983;221:236-40.

[233] Cochran B. The Molecular Action of Platelet Derived Growth Factor. *Adv. Cancer Res.* 1985;45;183-216.

[234] Evans HM, Long JA. The Effect of the Anterior Lobe Administered Intraperitoneally Upon Growth, Maturity and Oestrous cycles of the rat. *Anat. Rec.* 1921;21:62.

[235] Castagna M,Takai Y,Kaibuchi K,Sano K, Kikkawa U, Nishizuka Y. Direct Activation of Calcium Activated, Phospholipid Dependent Protein Kinase by Tumor Promoting Phorbol esters. *J. Biol. Chem.* 1982;257;7847-51.

[236] Cooper JA, Bowen-Pope DF,Raines E,Ross R,Hunter T. Similar Effects of Platelet Derived Growth Factor and Epidermal Growth Factor on the Phosphorylation of Tyrosine in Cellular Proteins. *Cell*. 1982;31:263-273.

[237] Schuldiner S and Rozengurt E. Na^+/H^+ Antiport in Swiss 3T3 cells:Mitogenic Stimulation Leads to Cytoplasmic Alkalinization. *Proc. Nat. Acad. Sci.* 1982;79:7778-82.

[238] Rozengurt E, Stroobant P, Waterfield MD, Deuel TF, Keehan M. Platelet Derived Growth Factor Elicits Cyclic AMP Accumulation in Swiss 3T3 Cells; *Role of Prostaglandin Production. cell.* 1983;34:265-72.

[239] Berridge MJ, Heslop JP, Irvine RF, Brown KD. Inositol Triphosphate Formation and Calcium Mobilization in Swiss 3T3 Cells in response to Platelet Derived Growth Factor. *Bio chem*. 1984;222:195-201.

[240] Gilman AG. G proteins and Dual Control of Adenylate Cyclase. *Cell*. 1984;36:577-579.

[241] Truneh A, Albert F, Goldstein P, Schmitt-Verhulst. Early Steps of Lymphocyte Activation By- passed by Synergy Between Calcium ionophores and Phorbol Ester. *Nature*. 1985;313:318-20.

[242] Balmain A, Gray J, Ponder B. The genetics and genomics of cancer. *Nat. Genet*. 2003;33(Suppl):238–44.

[243] Hartwell L. Defects in a cell cycle checkpoint may be responsible for the genomic instability of cancer cells. *Cell*. 1992;71:543–6.

[244] Raptis S, Bapat B. Genetic instability in human tumors. *EXS*. 2006:303–20.

[245] Renan MJ. How many mutations are required for tumorigenesis? Implications from human cancer data. *Mol. Carcinog*. 1993;7:139–46.

[246] Rajagopalan H, Nowak MA,Vogelstein B, et al. The significance of unstable chromosomes in colorectal cancer. *Nat. Rev. Cancer*. 2003;3:695–701.

[247] Kolodner RD. Mismatch repair: mechanisms and relationship to cancer susceptibility. *Trends Biochem. Sci.* 1995;20:397–401.

[248] Modrich P. Strand-specific mismatch repair in mammalian cells. *J. Biol. Chem*. 1997;272:24727–30.

[249] Lengauer C, Kinzler KW, Vogelstein B. Genetic instability in colorectal cancers. *Nature*. 1997;386:623–7.

[250] Raghavan SC, Lieber MR. DNA structures at chromosomal translocation sites. *Bioessays*. 2006;28:480–94.

[251] Gorringe KL, Chin SF, Pharoah P, et al. Evidence that both genetic instability and selection contribute to the accumulation of chromosome alterations in cancer. *Carcinogenesis*. 2005;26:923–30.

[252] Hughes S, Yoshimoto M, Beheshti B, et al. The use of whole genome amplification to study chromosomal changes in prostate cancer: insights into genome-wide signature of preneoplasia associated with cancer progression. *BMC Genomics*. 2006;7:65.

[253] Schrock E, Veldman T, Padilla-Nash H, et al. Spectral karyotyping refines cytogenetic diagnostics of constitutional chromosomal abnormalities. *Hum. Genet.* 1997;101:255–62.

[254] Speicher MR, Gwyn Ballard S, Ward DC. Karyotyping human chromosomes by combinatorial multi-fluor FISH. *Nat. Genet.* 1996;12:368–75.

[255] Chudoba I, Plesch A, Lorch T, et al. High resolution multicolor-banding: a new technique for refined FISH analysis of human chromosomes. *Cytogenet. Cell Genet.* 1999;84:156–60.

[256] Abeysinghe SS, Chuzhanova N, Krawczak M, et al. Translocation and gross deletion breakpoints in human inherited disease and cancer I: nucleotide composition and recombination-associated motifs. *Hum. Mutat.* 2003;22:229–44.

[257] Roy-Engel AM, Salem AH, Oyeniran OO, et al. Active Alu element "Atails": size does matter. *Genome Res.* 2002;12:1333–44.

[258] Batzer MA, Deininger PL. Alu repeats and human genomic diversity. *Nat. Rev. Genet.* 2002; 3: 370–9.

[259] Kolomietz E, Meyn MS, Pandita A, et al. The role of Alu repeat clusters as mediators of recurrent chromosomal aberrations in tumors. *Genes Chromosomes Cancer*. 2002;35:97–112.

[260] Edelmann L, Spiteri E, Koren K, et al. AT-rich palindromes mediate the constitutional t(11;22) translocation. *Am. J. Hum. Genet.* 2001;68:1–13.

[261] Braude I, Vukovic B, Prasad M, et al. Large scale copy number variation (CNV) at 14q12 is associated with the presence of genomic abnormalities in neoplasia. *BMC Genomics*. 2006;7:138.

[262] Vig BK, Sternes KL, Paweletz N. Centromere structure and function in neoplasia. *Cancer Genet. Cytogenet.* 1989;43:151–78.

[263] Mitelman F, Mertens F, Johansson B. A breakpoint map of recurrent chromosomal rearrangements in human neoplasia. *Nat. Genet.* 1997;15(Spec No):417–74.

[264] Grady DL, Ratliff RL, Robinson DL, et al. Highly conserved repetitive DNA sequences are present at human centromeres. *Proc. Natl. Acad. Sci. USA.* 1992;89:1695–9.

[265] Padilla-Nash HM, Heselmeyer-Haddad K, Wangsa D, et al. Jumping translocations are common in solid tumor cell lines and result in recurrent fusions of whole chromosome arms. *Genes Chromosomes Cancer.* 2001;30:349–63.

[266] Lejeune J, Maunoury C, Prieur M, et al. A jumping translocation (5p;15q), (8q;15q), and (12q;15q) (author's transl). *Ann. Genet.* 1979;22:210–3.

[267] Bayani J, Zielenska M, Pandita A, et al. Spectral karyotyping identifies recurrent complex rearrangements of chromosomes 8, 17, and 20 in osteosarcomas. *Genes Chromosomes Cancer.* 2003;36:7–16.

[268] Beheshti B, Karaskova J, Park PC, et al. Identification of a high frequency of chromosomal rearrangements in the centromeric regions of prostate cancer cell lines by sequential Giemsa banding and spectral karyotyping. *Mol. Diagn.* 2000;5:23–32.

[269] Vukovic B, Beheshti B, Park PC, et al. Correlating breakage-fusion-bridge events with the overall chromosomal instability and in vitro karyotypic evolution in prostate cancer. *Cytogenet Genome Res*; in press.

[270] Popescu NC. Genetic alterations in cancer as a result of breakage at fragile sites. *Cancer Lett.* 2003;192:1–17.

[271] Glover TW. Common fragile sites. *Cancer Lett.* 2006 ;232:4–12.

[272] Glover TW, Arlt MF, Casper AM, et al. Mechanisms of common fragile site instability. *Hum. Mol. Genet.* 2005;14(Spec No. 2):R197–205.

[273] Ferber MJ, Eilers P, Schuuring E, et al. Positioning of cervical carcinoma and Burkitt lymphoma translocation breakpoints with respect to the human papillomavirus integration cluster in FRA8C at 8q24. 13. *Cancer Genet. Cytogenet.* 2004;154:1–9.

[274] Holschneider CH, Baldwin RL, Tumber K, et al. The fragile histidine triad gene: a molecular link between cigarette smoking and cervical cancer. *Clin. Cancer Res.* 2005;11:5756–63.

[275] Yuan BZ, Keck-Waggoner C, Zimonjic DB, et al. Alterations of the FHIT gene in human hepatocellular carcinoma. *Cancer Res.* 2000;60:1049–53.

[276] Miller CT, Lin L, Casper AM, et al. Genomic amplification of MET with boundaries within fragile site FRA7G and upregulation of

METpathways in esophageal adenocarcinoma. *Oncogene.* 2006;25:409–18.

[277] Hoeijmakers JHJ. DNA damage, aging, and cancer. N Engl J Med. 2009; 1475e85:361.

[278] Mai S, Garini Y. Oncogenic remodeling of the three-dimensional organization of the interphase nucleus: c-Myc induces telomeric aggregates whose formation precedes chromosomal rearrangements. *Cell Cycle.* 2005;4:1327–31.

[279] Louis SF, Vermolen BJ, Garini Y, et al. c-Myc induces chromosomal rearrangements through telomere and chromosome remodelling in the interphase nucleus. *Proc. Natl. Acad. Sci. USA.* 2005;102:9613–8.

[280] Lim G, Karaskova J, Beheshti B, et al. An integrated mBAND and submegabase resolution tiling set (SMRT)CGHarray analysis of focal amplification, microdeletions, and ladder structures consistent with breakagefusion- bridge cycle events in osteosarcoma. *Genes Chromosomes Cancer.* 2005;42:392–403.

[281] Lim G, Karaskova J, Vukovic B, et al. Combined spectral karyotyping, multicolor banding, and microarray comparative genomic hybridization analysis provides a detailed characterization of complex structural chromosomal rearrangements associated with gene amplification in the osteosarcoma cell line MG-63. *Cancer Genet Cytogenet.* 2004;153:158–64.

[282] Pandita A, Zielenska M, Thorner P, et al. Application of comparative genomic hybridization, spectral karyotyping, and microarray analysis in the identification of subtype specific patterns of genomic changes in rhabdomyosarcoma. *Neoplasia.* 1999;1:262–75.

[283] Gisselsson D, Pettersson L, Hoglund M, et al. Chromosomal breakage fusion- bridge events cause genetic intratumor heterogeneity. *Proc. Natl. Acad. Sci. USA.* 2000;97:5357–62.

[284] Bayani J, Brenton JD, Macgregor PF, et al. Parallel analysis of sporadic primary ovarian carcinomas by spectral karyotyping, comparative genomic hybridization, and expression microarrays. *Cancer Res.* 2002;62:3466–76.

[285] Shuster MI, Han L, Le Beau MM, et al. A consistent pattern of RIN1 rearrangements in oral squamous cell carcinoma cell lines supports a breakage-fusion-bridge cycle model for 11q13 amplification. *Genes Chromosomes Cancer.* 2000;28:153–63.

[286] Shay JW,Wright WE. Telomeres and telomerase: implications for cancer and aging. *Radiat. Res.* 2001;155:188–93.

[287] Gisselsson D, Gorunova L, Hoglund M, et al. Telomere shortening and mitotic dysfunction generate cytogenetic heterogeneity in a subgroup of renal cell carcinomas. *Br. J. Cancer.* 2004;91:327–32.

[288] Gisselsson D, Hoglund M. Connecting mitotic instability and chromosome aberrations in cancer—can telomeres bridge the gap? *Semin Cancer Biol.* 2005;15:13–23.

[289] Gisselsson D. Chromosome instability in cancer: how, when, and why? *Adv Cancer Res.* 2003;87:1–29.

[290] FeinbergAP,Vogelstein B. Hypomethylation distinguishes genes of some human cancers from their normal counterparts. *Nature.* 1983;301:89–92.

[291] Chen RZ, Pettersson U, Beard C, et al. DNA hypomethylation leads to elevated mutation rates. *Nature.* 1998;395:89–93.

[292] Eden A, Gaudet F, Waghmare A, et al. Chromosomal instability and tumors promoted by DNA hypomethylation. *Science.* 2003;300:455.

[293] Egger G, Liang G, Aparicio A, Jones PA. Epigenetics in human disease and prospects for epigenetic therapy. *Nature.* 2004;429:457-63.

[294] Jones PA, Takai D. The role of DNA methylation in mammalian epigenetics. *Science.* 2001;293:1068-70.

[295] Herman JG, Baylin SB. Gene silencing in cancer in association with promoter hypermethylation. *N. Engl. J. Med.* 2003;349:2042-54.

[296] Ehrlich M. DNA methylation in cancer: too much, but also too little. *Oncogene.* 2002;21:5400-13.

[297] Kim H34. , Kwon YM, Kim JS, Lee H, Park JH, Shim YM, et al. Tumor-specific methylation in bronchial lavage for the early detection of non-small-cell lung cancer. *J. Clin. Oncol.* 2004;22:2363-70.

[298] Jenuwein T, Allis CD. Translating the histone code. Science. 2001;293:1074e80.

[299] Chi P, Allis CD, Wang GG. Covalent histone modifications: miswritten, misinterpreted and mis-erased in human cancers. *Nat. Rev. Cancer.* 2010;10:457e69.

[300] Ozda_g H, Teschendorff AE, Ahmed AA, Hyland SJ, Blenkiron C, Bobrow L, et al. Differential expression of selected histone modifier genes in human solid cancers. *BMC Genomics.* 2006;7:90e105.

[301] Nikoloski G, Langemeijer SM, Kuiper RP, Knops R, Massop M, To"nnissen ER, et al. Somatic mutations of the histone methyltransferase gene EZH2 in myelodysplastic syndromes. *Nat. Genet.* 2010;42:665e7.

[302] Hitchins MP, Wong JJ, Suthers G, Suter CM, Martin DI, Hawkins NJ, et al. Inheritance of a cancer-associated MLH1 germ-line epimutation. *N. Engl. J. Med.* 2007;356:697e705.

[303] Ligtenberg MJ, Kuiper RP, Chan TL, Goossens M, Hebeda KM,Voorendt M, et al. Heritable somatic methylation and inactivation of MSH2 in families with Lynch syndrome due to deletion of the 30 exons of TACSTD1. *Nat. Genet.* 2009;41:112e7.

[304] Baylin SB, Ohm JE. Epigenetic gene silencing in cancer—a mechanism for early oncogenic pathway addiction? *Nat. Rev. Cancer.* 2006;6:107–16.

[305] Fraga MF, Esteller M. Towards the human cancer epigenome: a first draft of histone modifications. *Cell Cycle.* 2005;4:1377–81.

[306] Lengauer C, Kinzler KW, Vogelstein B. DNA methylation and genetic instability in colorectal cancer cells. *Proc. Natl. Acad. Sci. USA.* 1997;94:2545–50.

[307] Rodriguez J, Frigola J, Vendrell E, et al. Chromosomal instability correlates with genome-wideDNA demethylation in human primary colorectal cancers. *Cancer Res.* 2006;66:8462–9468.

[308] Lindahl T, Wood RD. Quality control by DNA repair. *Science.* 1999;286:1897–905.

[309] Hoeijmakers JH. Genome maintenance mechanisms for preventing cancer. *Nature.* 2001;411:366–74.

[310] Khanna KK, Jackson SP. DNA double-strand breaks: signaling, repair and the cancer connection. *Nat. Genet.* 2001;27:247–54.

[311] Deng CX, Wang RH. Roles of BRCA1 in DNA damage repair: a link between development and cancer. *Hum. Mol. Genet.* 2003;12(Spec No. 1):R113–23.

[312] Meraldi P, Nigg EA. The centrosome cycle. *FEBS Lett.* 2002;521:9–13.

[313] Al-Romaih K, Bayani J, Vorobyova J, et al. Chromosomal instability in osteosarcoma and its association with centrosome abnormalities. *Cancer Genet Cytogenet.* 2003;144:91–9.

[314] Kops GJ, Weaver BA, Cleveland DW. On the road to cancer: aneuploidy and the mitotic checkpoint. *Nat. Rev. Cancer.* 2005;5:773–85.

[315] Warburg O. On the origin of cancer cells. *Science.* 1956;123:309-14.

[316] Dang CV, Semenza GL. Oncogenic alterations of metabolism. *Trends Biochem Sci.* 1999;24 :68-72.

[317] Kim JW, Tchernyshyov I, Semenza GL, Dang CV. HIF-1-117 mediated expression of pyruvate dehydrogenase kinase: a metabolic switch required for cellular adaptation to hypoxia. *Cell Metab.* 2006;3:177-85.

[318] Cuezva JM, Krajewska M, Heredia ML, Krajewski S, Kim GSH, Zapata JM, et al. The Bioenergetic signature of cancer: A marker of tumor progression. *Cancer Res.* 2002;62:6674-81.

[319] Mathupala SP, Ko YH, Pedersen PL. Hexokinase II: cancer's 126 double-edged sword acting as both facilitator and gatekeeper of malignancy when bound to mitochondria. *Oncogene.* 2006;25:4777-86.

[320] Hay N, Sonenberg N. Upstream and downstream of mTOR. *Genes Dev.* 2004;18:1926-45.

[321] Feng Z, Zhang H, Levine AJ, Jin S. The coordinate regulation of the p53 and mTOR pathways in cells. *Proc. Natl. Acad. Sci. USA.* 2005;102:8204-9.

[322] Blackburn EH. Telomere states and cell fates. *Nature.* 2000;408:53-6.

[323] Forsyth NR, Wright WE, Shay JW. Telomerase and differentiation in multicellular organisms: Turn it off, turn it on, and turn it off again. *Differentiation.* 2002;69:188-97.

[324] Shay JW, Bacchetti S. A survey of telomerase activity in. human cancer. *Eur. J. Cancer.* 1997;5:787-91.

[325] Shay JW, Wright WE. Telomerase therapeutics for cancer challenges and new directions. *Nat. Rev. Drug Disc.* 2006;5:577-84.

[326] Shay JW. Telomerase in cancer: diagnostic, prognostic and therapeutic implications. *Cancer J. Sci. Am.* 1998;(Suppl 1):S26-S34.

[327] Li H, Liu JP. Signalling on telomerase: a master switch in cell aging and immortalization. *Biogerontology.* 2002;3:109-18.

[328] Wong JMY, Kusdra L, Collins K. Subnuclear shuttling of human telomerase induced by transformation and DNA damage. *Nat. Cell Biol.* 2002;4:731-6.

[329] Levine AJ, Finlay CA, Hinds PW. P53 is a tumor suppressor gene. *Cell.* 2004;116:S67-9.

[330] Lowe SW, Cepero E, Evan G. Intrinsic tumor suppression. *Nature.* 2004;432:307-15.

[331] Erster S, Moll UM. Stress-induced p53 runs a direct mitochondrial death program: its role in physiologic and pathophysiologic stress responses in vivo. *Cell Cycle.* 2004;3:1492-5.

[332] Harms K, Nozell S, Chen X. The common and distinct target genes of the p53 family transcription factors. *Cell Mol. Life Sci.* 2004;61:822-42.

[333] Stoklosa T, Golab J. Prospects for p53-based cancer therapy. *Acta Biochim. Pol .* 2005; 52:321-8.

[334] Baselga J. Targeting tyrosine kinases in cancer: The second wave. *Science.* 2006;312:1175-8.

[335] Paul MK, Mukhopadhyay AK. Tyrosine kinase - Role and significance in cancer. *Int. J. Med. Sci.* 2004;1:101-15.

[336] Violette M, Helene RF. Role of histone N-terminal tails and their acetylation in nucleosome dynamics. *Mol. Cell Biol.* 2000;19 :7230-7.

[337] Munster PN, Troso S, Rosen N, Rifkind R, Marks PA. Richon VM. The histone deacetylase inhibitor suberoyanilide hydroxamic acid induces differentiation of human breast cancer cells. *Cancer Res.* 2001;61:8492-7.

[338] Kristeleit R, Stimson L, Workman P, Aherne W. Histone modification enzymes: novel targets for cancer drugs. *Expert Opin. Emerg. Drugs.* 2004;9:135-54.

[339] Lu KP, Suizu F, Zhou XZ, Finn G, Lam P, Wulf G. Targeting carcinogenesis: a role for the prolyl isomerase Pin1? *Mol. Carcinog.* 2006;45:397-402.

[340] Ryo A, Suizu F, Yoshida Y, Perrem K, Liou YC, Wulf G. et al. Regulation of NF-kappaB signaling by Pin1-catalyzed prolyl isomerization and ubiquitin-mediated proteolysis of p65/RelA. *Mol. Cell* 2003;12:1413-26.

[341] Ryo A, Liou YC, Wulf G, Nakamura N, Lee SW, Lu KP. Pin1 is an E2F target gene essential for the Neu/Ras-induced transformation of mammary epithelial cells. *Mol. Cell Biol.* 2002;22:5281-95.

[342] Basu A, Das M, Qanungo S, Fan XJ, DuBois G, Haldar S. Proteasomal degradation of human peptidyl prolyl isomerase pin1-pointing phosphor Bcl2 toward dephosphorylation. *Neoplasia.* 2002;4:218-27.

[343] Atchison FW, Capel B, Means AR. Pin1 regulates the timing of mammalian primordial germ cell proliferation. *Development.* 2003;130:3579-86.

[344] Rippmann FJ, Hobbie S, Daiber C, Guilliard B, Bauer M. Birk J, et al. Phosphorylation-dependent proline isomerization catalyzed by Pin1 is essential for tumor cell survival and entry into mitosis. *Cell Growth Differ.* 2000;11:409-16.

[345] Mathur R. Experimental studies on the modification of cellular responses to topoisomerase inhibitors in normal and transformed cell lines, Ph. D. thesis. University of Delhi; 2008.

[346] Mathur R, Suman S, Beaume N, Ali M, Bhatt AN, Chopra M, et al. Interaction, structural modification of topoisomerase IIα by peptidyl prolyl isomerase, Pin1: an in silico study. *Protein Pept. Lett.* 2010;17:151-63.

[347] Knudson AG. Two genetic hits(more or less) to cancer. *Nature.* 2001;1:157-62.

[348] Dwarakanath BS, Manogaran PS, Das S, Das BS, Jain V. Heterogeneity in DNA content and proliferative status of human brain tumors. *Indian J. Med. Res.* 1994;100:279-86.

[349] Whitfield ML,George LK, Grant GD,Perou CM. Common markers of proliferation. *Nat. Rev. Cancer.* 2006;6:99-106.

[350] Zarbo RJ, Nakhleh RE, Brown RD, Kubus JJ, Ma CK. Prognostic significance of DNA ploidy and proliferation in 309 colorectal carcinomas as determined by two-color multiparametric DNA flow cytometry. *Cancer.* 2000;79:2073-86.

[351] Bishop JM. The molecular genetics of cancer. Science (Wash. DC). 1987;235:305-11.

[352] Dunning AM, Healey CS, Pharoah PDP, Teare DM, Ponder BAJ, Easton DF. A systematic review of genetic polymorphisms and breast cancer risk. *Cancer Epidem Biomarkers Prevent.* 1999; 8:843-54.

[353] Toru H, Masaharu Y, Shinji T, Kazuaki C. Genetic polymorphisms and head and neck cancer risk. *Int. J. Oncol.* 2008;32:945-73.

[354] Nakayama H, Hibi K, Takase T, Yamazaki T, Kasai Y, Ito K, et al. Molecular detection of p16 promoter methylation in the serum of recurrent colorectal cancer patients. *Int. J. Cancer.* 2003;105:491-3.

[355] Belinsky SA. Gene-promoter hypermethylation as a biomarker in lung cancer. *Nat. Rev. Cancer.* 2004; 4:707-17.

[356] Esteller M, Garcia-Foncillas J, Andion E, Goodman SN, Hidalgo OF, Vanaclocha V, et al. Inactivation of the DNA-repair gene MGMT and the clinical response of gliomas to alkylating agents. *N. Engl. J. Med.* 2000; 343:1350-4.

[357] Dickson D. Wellcome funds cancer database. *Nature.* 1999;401:729.

[358] Hudson TJ, Anderson W, Artez A, Barker AD, Bell C, Bernabe´ RR, et al. International Cancer Genome Consortium. International network of cancer genome projects. *Nature.* 2010;464:993-8.

[359] Yan H, Parsons DW, Jin G, McLendon R, Rasheed BA, YuanW, et al. IDH1 and IDH2 mutations in gliomas. *N. Engl. J. Med.* 2009;360:765-73.

[360] Jones S, Hruban RH, Kamiyama M, Borges M, Zhang X, Parsons DW, et al. Exomic sequencing identifies PALB2 as a pancreatic cancer susceptibility gene. *Science.* 2009;324:217.

[361] Bernards R. It's diagnostics, stupid. *Cell.* 2010;141:13-7.

[362] Friend SH, Bernards R, Rogelj S, Weinberg RA, Rapaport JM, Albert DM, Dryja TP. A human DNA segment with properties of the gene that

predisposes   to   retinoblastoma   and   osteosarcoma.   *Nature.* 1986;323:643e6.

[363] Lichtenstein P, Holm NV, Verkasalo PK, Iliadou A, Kaprio J, Koskenvuo M, et al. Environmental and heritable factors in the causation of cancer: analyses of cohorts of twins from Sweden, Denmark, and Finland. *N. Engl. J. Med.* 2000;343:78-85.

[364] Tenesa A, Dunlop MG. New insights into the aetiology of colorectal cancer from genome-wide association studies. *Nat. Rev. Genet.* 2009;10:353-8.

[365] Kiemeney LA, Sulem P, Besenbacher S, Vermeulen SH, Sigurdsson A, Thorleifsson G, et al. A sequence variant at 4p16. 3 confers susceptibility to urinary bladder cancer. *Nat. Genet.* 2010;42:415-9.

[366] Tuupanen S, Turunen M, Lehtonen R, Hallikas O, Vanharanta S, Kivioja T, et al. The common colorectal cancer predisposition SNP rs6983267 at chromosome 8q24 confers potential to enhanced Wnt signaling. *Nat. Genet.* 2009;41:885-90.

[367] Pomerantz MM, Ahmadiyeh N, Jia L, Herman P, Verzi MP, Doddapaneni H, et al. The 8q24 cancer risk variant rs6983267 shows long-range interaction with MYC in colorectal cancer. *Nat. Genet.* 2009;41:882-4.

[368] Ignatiadis M, Xenidis N, Perraki M, Apostolaki S, Politaki E, Kafousi M, et al. Different prognostic value of Cytokeratin-19 mRNA positive circulating tumor cells according to estrogen receptor and HER2 status in early-stage breast cancer. *J. Clin. Oncol.* 2007;25:5194-202.

[369] Delys L, Detours V, Franc B, Thomas G, Bogdanova T, Tronko M, et al. Gene expression and the biological phenotype of papillary thyroid carcinomas. *Oncogene.* 2007;26:7894-903.

[370] Jeffery PL, Herington AC, Chopin LK. Ghrelin and a novel ghrelin isoform have potential autocrine/paracrine roles in hormone-dependent cancer. *Endocrine Abstracts.* 2003;6:38.

[371] Alison MR, Hunt T, Forbes SJ. Minichromosome maintenance (MCM) proteins may be pre-cancer markers. *Gut.* 2002;50:290-1.

[372] Forsburg SL. Eukaryotic MCM proteins: beyond replication initiation. *Microbiol. Mol. Biol. Rev.* 2004;68:109-31.

[373] Jakupciak JP, Wang W, Markowitz ME, Ally D, Coble M, Srivastava S, et al. Mitochondrial DNA as a cancer biomarker. *J. Mol. Diagn.* 2005;7:258-67.

[374] Maitra A, Cohen Y, Gillespie SE, Mambo E, Fukushima N, Hoque MO, et al. The Human MitoChip: a highthroughput sequencing microarray for mitochondrial mutation detection. *Genome Res.* 2004;14:812-9.

[375] Kim MM, Clinger JD, Masayesva BG, Ha PK, Zahurak ML,Westra WH, et al. Mitochondrial DNA quantity increases with histopathologic grade in premalignant and malignant head and neck lesions. *Cancer Res.* 2004;10:8512-5.

[376] Lièvre A, Blons H, Houllier AM, Laccourreye O, Brasnu D, Beaune P, et al. Clinicopathological significance of mitochondrial D-Loop mutations in head and neck carcinoma. *Br. J. Cancer.* 2006;94:692-7.

[377] Fliss MS, Usadel H, Caballero OL, Li WU, Buta MR, Eleff SM, et al. Facile detection of mitochondrial DNA mutations in tumors and bodily fluids. *Science.* 2000;287:2017-9.

[378] Kwok PY. Methods for genotyping single nucleotide polymorphisms. *Annu. Rev. Genomics Hum. Genet.* 2001;2:235-58.

[379] Maris JM,Hii G,Gelfand CA,Varde S,White PS,Rappaport E, et al. Region specific detection of neuroblastoma loss of heterozygosity at multiple loci simultaneously using a SNP-based tag-array platform. *Genome Res.* 2005;15:1168-76.

[380] Xu P, Vernooy SY, Guo M, et al. The Drosophila microRNA Mir-14 suppresses cell death and is required for normal fat metabolism. *Curr. Biol.* 2003;13:790-5.

[381] Calin G A, Dumitru C D, Shimizu M, et al. Frequent deletions and down-regulation of micro-RNA genes miR15 and miR16 at 13q14 in chronic lymphocytic leukemia. *PNAS.* 2002;99:15524-29.

[382] Calin GA, Sevignani C, Dumitru CD, et al. Human microRNA genes are frequently located at fragile sites and genomic regions involved in cancers. *PNAS.* 2004;101:2999-3004.

[383] Calin GA, Croce CM. MicroRNA signatures in human cancers. *Nat. Rev. Cancer.* 2006;6:857-66.

[384] Esquela-Kerscher A, Slack FJ. Oncomirs: MicroRNAs with a role in cancer. *Nat. Rev. Cancer.* 2006;6:259-69.

[385] Lu J, Getz G, Miska EA. MicroRNA expression profiles classify human cancers. *Nature.* 2005;435:834-838.

[386] Li J, Smyth P, Flavin R, et al. Comparison of miRNA expression patterns using total RNA extracted from matched samples of formalin-fixed paraffin-embedded (FFPE) cells and snap-frozen cells. *BMC Biotechnol.* 2007;7:36.

[387] Yan X J, Albesiano E, Xi Y,et al. Systematic analysis of microRNA expression of RNA extracted from fresh frozen and formalin-fixed paraffin-embedded samples. *RNA*. 2007;13:1668-74.

[388] Pineau P, Volinia S, McJunkin K, Marchio A, Battiston C, Terris B, et al. Mir-221 overexpression contributes to liver tumorigenesis. *Proc. Natl. Acad. Sci. U S A*. 2010;107:264-9.

[389] Volinia S, Calin GA, Liu CG, Ambs S, Cimmino A, Petrocca F, et al. A microRNA expression signature of human solid tumors defines cancer gene targets. *Proc. Natl. Acad. Sci. U S A*. 2006;103:2257–61.

[390] Tomimaru Y, Eguchi H, Nagano H, Wada H, Tomokuni A, Kobayashi S, et al. MicroRNA-21 induces resistance to the anti-tumor effect of interferon-alpha/5- fluorouracil in hepatocellular carcinoma cells. *Br. J. Cancer*. 2010;103:1617–26.

[391] Karakatsanis A, Papaconstantinou I, Gazouli M, Lyberopoulou A, Polymeneas G, Voros D. Expression of microRNAs, miR-21,miR-31,miR-122,miR-145, miR-146a,miR-200c, miR-221, miR-222, and miR-223 in patients with hepatocellular carcinoma or intrahepatic cholangiocarcinoma and its prognostic significance. Mol Carcinog in press, http://dx. doi. org/10. 1002/mc. 21864.

[392] Tomimaru Y, Eguchi H, Nagano H, Wada H, Kobayashi S, Marubashi S, et al. Circulating microRNA-21 as a novel biomarker for hepatocellular carcinoma. *J. Hepatol*. 2012;56:167–75.

[393] Li W, Xie L, He X, Li J, Tu K, Wei L, et al. Diagnostic and prognostic implications of microRNAs in human hepatocellular carcinoma. *Int. J. Cancer*. 2008;123:1616–22.

[394] Park JK, Kogure T, Nuovo GJ, Jiang J, He L, Kim JH, et al. miR-221 silencing blocks hepatocellular carcinoma and promotes survival. *Cancer Res*. 2011;71:7608–16.

[395] Ding J, Huang S,Wu S, Zhao Y, Liang L, Yan M, et al. Gain of miR-151 on chromosome 8q24. 3 facilitates tumor cell migration and spreading through downregulating RhoGDIA. *Nat. Cell Biol*. 2010;12:390–9.

[396] Liu S, Guo W, Shi J, Li N, Yu X, Xue J, et al. MicroRNA-135a contributes to the development of portal vein tumor thrombus by promoting metastasis in hepatocellular carcinoma. *J. Hepatol* . 2012;56:389–96.

[397] Xu T, Zhu Y, Wei QK, Yuan Y, Zhou F, Ge YY, et al. A functional polymorphism in the miR-146a gene is associated with the risk for hepatocellular carcinoma. *Carcinogenesis*. 2008;29:2126–31.

[398] Pandey DP, Picard D. miR-22 inhibits estrogen signaling by directly targeting the estrogen receptor alpha mRNA. *Mol. Cell Biol.* 2009;29:3783–90.

[399] Xiong J, Yu D, Wei N, Fu H, Cai T, Huang Y, et al. An estrogen receptor alpha suppressor, microRNA-22, is downregulated in estrogen receptor alpha-positive human breast cancer cell lines and clinical samples. *FEBS J.* 2010;277:1684–94.

[400] Zhang J, Yang Y, Yang T, Liu Y, Li A, Fu S, et al. MicroRNA-22, downregulated in hepatocellular carcinoma and correlated with prognosis, suppresses cell proliferation and tumorigenicity. *Br. J. Cancer.* 2010;103:1215–20.

[401] Ji J, Zhao L, Budhu A, Forgues M, Jia HL, Qin LX, et al. Let-7g targets collagen type I alpha2 and inhibits cell migration in hepatocellular carcinoma. *J. Hepatol.* 2010;52:690–7.

[402] Webster RJ, Giles KM, Price KJ, Zhang PM, Mattick JS, Leedman PJ. Regulation of epidermal growth factor receptor signaling in human cancer cells by microRNA-7. *J. Biol. Chem.* 2009;284:5731–41.

[403] Fang YX, Xue JL, Shen Q, Chen J, Tian L. miR-7 inhibits tumor growth and metastasis by targeting the PI3K/Akt pathway in hepatocellular carcinoma. *Hepatology.* 2012;55(6):1852–62.

[404] Tomokuni A, Eguchi H, Tomimaru Y, Wada H, Kawamoto K, Kobayashi S, et al. miR-146a suppresses the sensitivity to interferon-alpha in hepatocellular carcinoma cells. *Biochem. Biophys. Res. Commun.* 2011;414:675–80.

[405] Patel T. Cholangiocarcinoma. *Nat. Clin. Pract. Gastroenterol. Hepatol.* 2006;3:33–42.

[406] Meng F, Henson R, Lang M, Wehbe H, Maheshwari S, Mendell JT, et al. Involvement of human micro-RNA in growth and response to chemotherapy in human cholangiocarcinoma cell lines. *Gastroenterology.* 2006;130:2113–29.

[407] He Q, Cai L, Shuai L, Li D, Wang C, Liu Y, et al. Ars2 is overexpressed in human cholangiocarcinomas and its depletion increases PTEN and PDCD4 by decreasing microRNA-21. Mol Carcinog in press, http://dx. doi. org/10. 1002/mc. 21859.

[408] Meng F, Henson R, Wehbe-Janek H, Ghoshal K, Jacob ST, Patel T. MicroRNA-21 regulates expression of the PTEN tumor suppressor gene in human hepatocellular cancer. *Gastroenterology.* 2007;133:647–58.

[409] Guo J, Miao Y, Xiao B, Huan R, Jiang Z, Meng D, et al. Differential expression of microRNA species in human gastric cancer versus nontumorous tissues. *J. Gastroenterol. Hepatol.* 2009;24:652–7.

[410] Hao J, Zhang S, Zhou Y, Liu C, Hu X, Shao C. MicroRNA 421 suppresses DPC4/Smad4 in pancreatic cancer. *Biochem. Biophys. Res. Commun.* 2011;406:552–7.

[411] Zhong XY, Yu JH, Zhang WG, Wang ZD, Dong Q, Tai S, et al. MicroRNA-421 functions as an oncogenic miRNA in biliary tract cancer through down-regulating farnesoidxreceptor expression. *Gene.* 2012;493:44–51.

[412] Razumilava N, Bronk SF, Smoot RL, Fingas CD, Werneburg NW, Roberts LR, et al. miR-25 targets TNF-related apoptosis inducing ligand (trail) death receptor-4 and promotes apoptosis resistance in cholangiocarcinoma. *Hepatology.* 2012;55: 465–75.

[413] Olaru AV, Ghiaur G, Yamanaka S, Luvsanjav D, An F, Popescu I, et al. MicroRNA down-regulated in human cholangiocarcinoma control cell cycle through multiple targets involved in the G1/S checkpoint. *Hepatology.* 2011;54:2089–98.

[414] Meng F, Wehbe-Janek H, Henson R, Smith H, Patel T. Epigenetic regulation of microRNA-370 by interleukin-6 in malignant human cholangiocytes. *Oncogene.* 2008;27:378–86.

[415] Chen Y, Gao W, Luo J, Tian R, Sun H, Zou S. Methyl-CpG binding protein MBD2 is implicated in methylation-mediated suppression of miR-373 in hilar cholangiocarcinoma. *Oncol. Rep.* 2011;25:443–51.

[416] Chen Y, Luo J, Tian R, Sun H, Zou S. miR-373 negatively regulates methyl-CpG-binding domain protein 2 (MBD2) in hilar cholangiocarcinoma. *Dig. Dis. Sci.* 2011;56:1693–701.

[417] Chen X, Ba Y, Ma L, et al. Characterization of microRNAs in serum: a novel class of biomarkers for diagnosis of cancer and other diseases. *Cell Res.* 2008;18:997-1006.

[418] Lawrie CH, Gal S, Dunlop HM, et al. Detection of elevated levels of tumor-associated microRNAS in serum of patients with diffuse large B-cell lymphoma. *Br. J. Heametol.* 2008;141:672-675.

[419] Mitchell PS, Parkin RK, Kroh EM, et al. Circulating microRNAs as stable blood-based markers for cancer detection. *PNAS.* 2008;105:10513-18.

[420] Calin GA, Ferracin M, Cimmino A, et al . A microRNA signature associated with prognosis and progression in chronic lymphocytic leukemia. *N. Engl. J. Med.* 2005;353:1793- 1801.

[421] Yanaihara N, Caplen N, Bowman E, et al. Unique microRNA molecular profiles in lung cancer diagnosis and prognosis. *Cancer Cell.* 2006;9:189-198.

[422] Lebanony D, Benjamin H, Gilad S, et al. Diagnostic assay based on - miR-205 expression distinguishes squamous from nonsquamous non-small-cell lung carcinoma. *J. Clin. Oncol.* 2009;27:2030-37.

[423] Raponi M, Dossey L, Jatkoe T, et al. MicroRNA classifiers for predicting prognosis of squamous cell lung cancer. *Cancer Res.* 2009;69:5776-83.

[424] Rabinowits G, Gerçel-Taylor C, Day JM, et al. Exosomal microRNA: a diagnostic marker for lung cancer. *Clin. Lung Cancer.* 2009;10:42-46.

[425] Rosell R, Wei J, Taron M. Circulating microRNA signature of tumor-derived exosomes for early diagnosis of non-small cell lung cancer. *Clin. Lung Cancer.* 2009;10:8-9.

[426] Nam EJ, Yoon H, Kim SW, et al. MicroRNA expression profiles in serous ovarian carcinoma. *Clin. Cancer Res.* 2008;14:2690-85.

[427] Laios A, O'Toole S, Flavin R, et al. Potential role of miR-9 and miR-233 in recurrent ovarian cancer. *Molecular Cancer.* 2008;7:35.

[428] Resnick KE, Alder H, Hagan JP, et al. The detection of differentially expressed microRNAs from the serum of ovarian cancer patients using a novel real-time PCR platform. *Genecol. oncol.* 2009;1121:5-59.

[429] Ng EKO, Chong WWS, Jin H, et al. Differential expression of microRNAs in plasma of colorectal cancer patients: a potential marker for colorectal cancer screening. *Gut.* 2009;58:1375-1381.

[430] Schetter AJ, Harris CC. Plasma microRNAs: a potential biomarker for colorectal cancer? *Gut.* 2009;58:1318-9.

[431] Feber A, Xi L, Luketich JD, et al. MicroRNA expression profiles of esophageal cancer. *J. Thorac. Cardiovasc. Surg.* 2008;135:255-260.

[432] Mathe EA, Nauyen GH, Bowman ED, et al. MicroRNA expression in squamous cell carcinoma and adenocarcinoma of the esophagus: associations with survival. Clin *Cancer Res.* 2009;15:6192-6200.

[433] Esteller M, Garcia-Foncillas J, Andion E, Goodman SN, Hidalgo OF, Vanaclocha V, *et al.* Inactivation of the DNA-repair gene *MGMT* and the clinical response of gliomas to alkylating agents. *N. Engl. J. Med.* 2000;343:1350-4.

[434] Ring A, Smith IE, Dowsett M. Circulating tumor cells in breast cancer. *Lancet Oncol.* 2004;5:79-88.

[435] Cristofanilli M, Budd GT, Ellis MJ, Stopeck A, Matera J, Miller MC, et al. Circulating tumor cells predict progression free survival and overall survival in metastatic breast cancer. *N. Engl. J. Med.* 2004;351:781-91.

[436] Shaffer DR, Leversha MA, Danila DC, Lin O, Gonzalez-Espinoza R, Gu B, et al. Circulating tumor cell analysis in patients with progressive castration resistant prostate cancer. *Clin. Cancer Res.* 2007;13:2023-9.

[437] Cristofanilli M, Budd GT, Ellis MJ, Stopeck A, Matera J. Miller MC, et al. Presence of circulating tumor cells (CTC) in metastatic breast cancer (MBC) predicts rapid progression and poor prognosis. *J. Clin. Oncol.* 2005;23:524.

[438] Sakaguchi S. Naturally arising CD4+ regulatory T cells for immunologic self-tolerance and negative control of immune responses. *Ann. Rev. Immunol.* 2004; 22:531-62.

[439] Maloy KJ, Powrie F. Regulatory T cells in the control of immune pathology. *Nat. Immunol.* 2001;2:816-22.

[440] Liyanage UK, Moore TT, Joo H-G, Tanaka Y, Herrmann V,Doherty G, et al. Prevalence of regulatory T cells is increased in peripheral blood and tumor microenvironment of patients with pancreas or breast adenocarcinoma. *J. Immunol.* 2002;169:2756-61.

[441] Woo EY, Yeh H, Chu CS, Schlienger K, Carroll RG, Riley JL, et al. Regulatory T cells from lung cancer patients directly inhibit autologous T cell proliferation. *J. Immunol.* 2002;168:4272-6.

[442] Ormandy LA, Hillemann T,Wedemeyer H, Manns MP, Greten TF, Korangy F. Increased populations of regulatory T cells in peripheral blood of patients with hepatocellular carcinoma. *Cancer Res.* 2005;65 :2457-64.

[443] Curiel T, Coukos G, Zou L,Alvarez X, Cheng P, Mottram P, et al. Specific recruitment of regulatory T cells in ovarian carcinoma fosters immune privilege and predicts reduced survival. *Nature Med.* 2004;10:942-9.

[444] Schreiber TH. The use of FoxP3 as a biomarker and prognostic factor for malignant human tumors. *Cancer Epidemiol Biomarkers Prevent.* 2007;16:1931-4.

[445] Clarke MF, Fuller M. Stem cells and cancer: two faces of eve. *Cell.* 2006;124:1111-5.

[446] Takaishi S, Okumura T, Wang TC. Gastric cancer stem cells. *J. Clin. Oncol.* 2008;26:2876-82

[447] Chen CJ, Wang LY, Yu MW. Epidemiology of hepatitis B virus infection in the Asia-Pacific region. *J. Gastroenterol. Hepatol.* 2000; 15(Suppl): E3-E6.

[448] Garland SM. Can cervical cancer be eradicated by prophylactic HPV vaccination? Challenges to vaccine implementation. *Indian J. Med. Res.* 2009;130:311-21.

[449] Shukla S, Bharti AC, Mahata S, Hussain S, Kumar R, Hedau S, et al. Infection of human papillomaviruses in cancers of different human organ sites. *Indian J. Med. Res.* 2009;130:222-33.

[450] Thompson MP, Kurzrock R. Epstein-Barr virus and cancer. *Clin. Cancer Res.* 2004;10:803-21.

[451] Papsidero LD, Wang MC, Valenzuela LA, Murphy GP, Chu TMA. Prostate antigen in sera of prostatic cancer patients. *Cancer Res.* 1980;40:2428-32.

[452] Thompson I, Pauler D, Goodman P, Tangen C, Lucia M, Parnes H, et al. Prevalence of prostate cancer among men with a prostate-specific antigen level < or =4. 0 ng per milliliter. *N. Engl. J. Med.* 2004;350:2239-46.

[453] Abelev GI. Alpha-fetoprotein in ontogenesis and its association with malignant tumors. *Adv. Cancer Res.* 1971;14:295-358.

[454] Paul SB, Gulati MS, Sreenivas V, Madan K, Gupta AK, Mukhopadhyay S, et al. Evaluating patients with cirrhosis for hepatocellular carcinoma: value of clinical symptomatology, imaging and alpha-fetoprotein. *Oncology.* 2007;72:117-23.

[455] Ball D, Rose E, Alpert E. Alpha-fetoprotein levels in normal adults. *Am. J. Med. Sci.* 1992;303:157-9.

[456] O'Brien TJ, Tanimoto H, Konishi I, Gee M. More than 15years of CA 125: what is known about the antigen, its structure and its function. *Int. J. Biol. Markers.* 1998;13:188-95.

[457] Bast RC Jr, Urban N, Shridhar V, Smith D, Zhang Z, Skates S,et al. Early detection of ovarian cancer: promise and reality. *Cancer Treat Res.* 2002;107:61-97.

[458] Meden H, Fattahi-Meibodi A. CA 125 in benign gynecological conditions. *Int. J. Biol. Markers.* 1998;13:231-7.

[459] Bast RC, Xu FJ, Yu YH, Barnhill S, Zhang Z, Mills GB. CA125: the past and the future. *Int. J. Biol. Markers.* 1998;13:179-87.

[460] Grover S, Koh H, Weideman P, Quinn MA. The effect of the menstrual cycle on serum CA 125 levels: a population study. *Am. J. Obstet. Gynecol.* 1992;167:1379-81.

[461] Van der Burg ME, Lammes FB, van Putten WL, Stoter G. Ovarian cancer: the prognostic value of the serum half-life of CA125 during induction chemotherapy. *Gynecol. Oncol.* 1988;30:307-12.

[462] Duffy MJ. CA 15-3 and related mucins as circulating markers in breast cancer. *Ann. Clin. Biochem.* 1999;36:579-86.

[463] Shering S, Sherry F, McDermott E, Higgins NO, Duffy MJ. Preoperative CA 15-3 concentrations predict outcome in breast cancer. *Cancer.* 1998;83:2521-7.

[464] Koprowski H, Steplewski Z, Mitchell K, Herlyn M, Herlyn D, Fuhrer P. Colorectal carcinoma antigens detected by hybridoma antibodies. *Somatic Cell Genet.* 1979;5:957-71.

[465] Casetta G, Piana P, Cavallini A, Vottero M, Tizzani A. Urinary levels of tumor associated antigens (CA 19-9, TPA and CEA) in patients with neoplastic and non-neoplastic urothelial abnormalities. *Br. J. Urol.* 1993;72:60-4.

[466] Duffy MJ. CA 19-9 as a marker for gastrointestinal cancers:a review. *Ann. Clin. Biochem.* 1998;35:364-70.

[467] Uygur-Bayramicli O, Dabak R, Orbay E, Dolapcioglu C, Sargin M, Kilicoglu G, et al. Type 2 diabetes mellitus and CA 19-9 levels. *World J. Gastroenterol.* 2007;13:5357-9.

[468] Alaoui-Jamali MA, Xu Y. Proteomic technology for biomarker profiling in cancer: an update. *J. Zhejiang Univ. Sci. B.* 2006;7:411-20.

[469] Khan MS, Chaouachi K, Mahmood R. Hookah. smoking and cancer: carcinoembryonic antigen (CEA) levels in exclusive/ever hookah smokers. *Harm. Reduction J.* 2008; 5:19.

[470] Wang JY, Lu CY, Chu KS, Ma CJ, Wu DC, Tsai HL, et al. Prognostic significance of pre- and postoperative serum carcinoembryonic antigen levels in patients with colorectal cancer. *Eur. Surg. Res.* 2007; 39:245-50.

[471] Cole LA. Immunoassay of human chorionic gonadotropin, its free subunits, and metabolites. *Clin. Chem.* 1997;43:2233-43.

[472] Kurtzman J, Wilson H, Rao CV. A proposed role for hCG in clinical obstetrics. *Sem. Reprod. Med.* 2001;19:63-8.

[473] Mazzaferri EL, Robbins RJ, Spencer CA, Braverman LE, Pacini F, Wartofsky L, et al. A consensus report of the role of serum thyroglobulin as a monitoring method for low-risk patients with papillary thyroid carcinoma. *J. Clin. Endocrinol. Metab.* 2003;88:1433-41.

[474] Pacini F, Fugazzola L, Lippi F, Ceccarelli C, Centoni R. Miccoli P, et al. Detection of thyroglobulin in fine needle aspirates of non thyroidal neck

masses: a clue to the diagnosis of metastatic differentiated thyroid cancer. *J. Clin. Endocrinol. Metab.* 1992;74:1401-4.

[475] Baloch Z, Carayon P, Conte-Devolx B, Demers LM, Rasmussen U, Henry JF, et al. Laboratory medicine practice guidelines. Laboratory support for the diagnosis and monitoring of thyroid disease. *Thyroid.* 2003;13:3-126.

[476] Kloos RT, Mazzaferri EL. A single recombinant human thyrotropin-stimulated serum thyroglobulin measurement predicts differentiated thyroid carcinoma metastases three to five years later. *J. Clin. Endocrinol. Metab.* 2005;90:5047-57.

[477] Voellmy R. On mechanisms that control heat shock transcription factor activity in metazoan cells. *Cell Stress Chaperones.* 2004; 9:122-33.

[478] Pinashi-Kimhi O, Michalowitz D, Ben-Zeev A, Oren M. Specific interaction between the p53 cellular tumor antigen and major heat shock proteins. *Nature.* 1986;320:182-5.

[479] Daniels GA, Sanchez-Perez L, Diaz RM, Kottke T, Thompson J, Lia M, et al. A simple method to cure established tumors by inflammatory killing of normal cells. *Nat. Biotechnol.* 2004;22:1125-32.

[480] Cioccal DR, Stuart K. Calderwood Heat shock proteins in cancer: diagnostic, prognostic, predictive, and treatment implications. *Cell Stress Chaperones.* 2005;10:86-103.

[481] Hu S, Yu T, Xie Y, Yang Y, Li Y, Zhou X, et al. Discovery of oral fluid biomarkers for human oral cancer by mass spectrometry. *Cancer Genomics Proteomics.* 2007:4:55-64.

[482] Hu S, Arellano M, Boontheung P, Wang J, Zhou H, Jiang J, et al. Salivary proteomics for oral cancer biomarker discovery. *Clin Cancer Res.* 2008;14:6246-52.

[483] Schlichtholz B, Legros Y, Gillet D, Gaillard C, Marty M, Lane D, et al. The immune response to p53 in breast cancer patients is directed against immunosdominant epitopes unrelated to the mutational hot spot. *Cancer Res.* 1992;52:6380-84.

[484] Trivers GE, De Benedetti VM, Cawley HL, Caron G, Harrington AM, Bennett WP, et al. Anti-p53 antibodies in sera from patients with chronic obstructive pulmonary disease can predate a diagnosis of cancer. *Clin. Cancer Res.* 1996;2:1767-1775.

[485] Fernandez Madrid F. Autoantibodies in breast cancer sera: candidate biomarkers and reporters of tumorigenesis. *Cancer Lett.* 2005;230:187-198.

[486] Jager E, Stockert E, Zidianakis Z, Chen YT, Karbach J, Jager D, et al. Humoral immune responses of cancer patients against "Cancer-Testis" antigen NY-ESO-1:correlation with clinical events. *Int. J. Cancer.* 1999;84:506-510.

[487] Disis ML, Knutson KL, Schiffman K, Rinn K, McNeel DG. Pre existent immunity to the HER-2/neu oncogenic protein in patients with HER-2/neu overexpressing breast and ovarian cancer. *Breast Cancer Res. Treat.* 2000;62:245-252.

[488] Oyama T, Osaki T,Baba T,et al. Molecular Genetic Tumor Markers in Non- small cell lung cancer. *Anticancer Res.* 2005;25:1193-1196.

[489] Cuperlovic-Culf M, Belacel N, Ouellette RJ. Determination of Tumor Marker genes from gene expression data. *Drug Discov. Today.* 2005;10:429-437.

In: Human Genome   ISBN: 978-1-62808-803-8
Editor: Tomeo Caccavelli   © 2013 Nova Science Publishers, Inc.

Chapter II

Evolution of Human Genome Analysis: Impact on Diseases Diagnosis and Molecular Diagnostic Labs

Julie Gauthier[1], Isabelle Thiffault[1],*
Virginie Dormoy-Raclet[1] and Guy A. Rouleau[1,2]
[1]Medical Biological Unit, Molecular Diagnostic Laboratory, Sainte-Justine
University Hospital Center, Montreal, QC, Canada
[2]Montreal Neurological Institute and Hospital,
Montréal (Québec), Canada

Abstract

The advent of massively parallel sequencing has changed the
interrogation process of the human genome and now provides a high
resolution and global view of the genome which is beyond research
applications. Together with powerful bioinformatics tools, these next
generation sequencing technologies have revolutionized fundamental
research and have important consequences for clinically actionable tests,
diagnosis and treatment of rare diseases and cancers. Today, molecular
testing is commonly used to confirm clinical diagnosis of specific

---

[*] Corresponding author: julie. gauthier@recherche-ste-justine. qc. ca.

diseases; it requires that a clinician specify the gene or mutation to test and, in return, will receive information only about this sequence. Despite relative successes, a large number of patients receive no accurate diagnosis, even after many expensive molecular investigations. A clear paradigm shift has taken place in the health network with the introduction of the exome sequencing in molecular diagnostic lab. In this chapter, the impact of the implementation of high throughput sequencing technologies on molecular diagnosis and on the practice of medicine, with an emphasis in paediatrics, is reviewed. We compared well-established genetic tests, using examples from our molecular diagnostic lab, to the recent exome sequencing applications. The genetic tests can fall into three main categories: 1) Mendelian Single Gene Disorder tests that include targeted mutation and targeted gene approaches 2) Genetic Disease Panels which are composed of a few to a dozen genes and 3) Exome or Genome approaches, which interrogate either the entire coding sequences of the 22,333 human genes or the entire human genome. For each of these categories, advantages and limitations are discussed. We devoted a section on the future of molecular diagnosis and discuss which tests will subsist and which one may be soon abandonned. Massively parallel sequencing is transforming the molecular diagnostic field: it offers personalized genetic tests and generates new ethical challenges. Important questions like incidental findings and possible forms of discrimination are addressed. Finally, we conclude with a section on the future directions surrounding the application of these multimodal molecular approaches in general and their putative applications in neonatal intensive care units.

1. Increased Human Genome Knowledge Through Progress in Sequencing Technology

Sequencing the DNA of an organism is feasible since 1977 when the "dideoxy" chain-termination method for sequencing DNA molecules was introduced by Frederick Sanger and colleagues [1]. This technique allows for the sequencing of a single DNA fragment up to 1000 base pairs long. The original method used radiolabeled dNTP and the reading was done manually. Sanger sequencing has been the basis of several major gene discoveries transforming the field of molecular biology. For example, before DNA sequencing, proteins were sequenced directly, this is a laborious technique. Now, this is easily accomplished by sequencing a cDNA and translating the

DNA sequence into the amino acid sequence of the protein. In the early 1990s, DNA sequencing was automated using a 4-channel capillary approach (basically the current Life Technologies ABI 3730 system) and ddNTP labeled with different color fluorescent dyes which allow fast DNA sequencing, with hundreds of fragments simultaneously. These sequencing systems were the first generation of sequencing technologies. Their high-throughput yield allows a single lab to sequence millions of base pairs, compared to thousands before their introduction. Commonly called the Sanger method, this sequencing technique is the gold standard in research and clinical diagnostic laboratories for genetic testing. This advance in technology led to science's greatest achievement, the Human Genome Project. This project aimed to determine the sequence of the 3 billion nucleotide base pairs that constitute the human genome and to identify all 22,333 genes [2] in human DNA [3]. This 13-year project gave rise to the first draft of the human genome sequence in 2001. Decoding the human genome has opened unprecedented new avenues in research and increased our understanding of human health. Information from the human genome sequence enables us to understand how the genetic information drives the development and function of the human body. More importantly, the Human Genome Project accelerated the exploration of genetic variations predisposing to disease and thus revolutionizes medical practice and biological research. Since then, the availability of genomic information (gene and chromosome structure, polymorphisms, disease causing mutations, etc.), has continuously increased.

Sanger sequencing helped the discovery of new genetic mechanisms. For example, by sequencing more than 400 synaptic genes in affected probands with sporadic schizophrenia or autism (that is, with no history of psychiatric disorders in the parents or the extended family), we and others have observed a significant excess of potentially deleterious *de novo* mutations in affected individuals [4]. A similar observation was done in intellectual disability cases [5].

The most recent DNA sequencing development is the advent of massively parallel sequencing platforms leading to the "next generation sequencing" technologies. The development of ultra-high throughput sequencing technologies pushed forward at an unprecedented speed the identification of DNA variations associated with diseases. Ultrasequencing platforms can be combined with microarray technologies to capture and amplify the functional genome that encodes proteins, also known as the "exome", or to capture more comprehensively a subset of related genes. As technology progressed, next-generation DNA sequencing techniques have improved significantly and are now much faster, and to some extent less expensive than Sanger sequencing. What took over 10 years like the Human Genome Project can now be done in less than a month. It's now feasible to decode multiple human genomes at once. All of these progresses gave birth to larger scale genetics.

A remarkable increase of genetic and genomic data occurred in the last decade notably through the introduction of microarrays and next-generation sequencing. In fact, human genome array-comparative genomic hybridization and SNP arrays were developed to detect genetic aberrations or copy number variants (CNVs) at a higher resolution than traditional karyotyping. Comparative genomic hybridization, also called array CGH, is now currently used in diagnostic labs to identify small chromosomal deletions and duplicated regions. For example, conventional karyotyping has a resolution of 5-10 million bases compare to 100 000 for array-CGH. Therefore, submicroscopic chromosomal alterations can now be detected. These technological progresses lead researchers to major discoveries in the understanding of genetic mechanisms of different neuropsychiatric conditions such as intellectual disabilities, autism and schizophrenia. De Vries et al. used array based comparative genomic hybridization (array CGH) to study 100 patients with unexplained intellectual disability. They were able to detect a potential causative chromosomal anomaly in 10% of their patients which represents a diagnostic yield of at least twice as high as that of conventional method [7]. In addition to the identification of the causative genetic factors implicated in pediatric and childhood complex disorders, microarrays initiated the discovery of new genetic mechanisms for a number of complex disorders, namely autism, intellectual disability and schizophrenia. Using these technologies, studies have highlighted the involvement of rare (<1% frequency) point mutations and CNVs in the genetic etiology of autism, schizophrenia, intellectual disability, attention deficit disorder and other disorders [6-9].

Molecular diagnostics has become an essential tool for the evaluation and management of patients, particularly of children with genetic diseases. According to the Online Mendelian Inheritance in Man (OMIM), an online catalog of human genes and genetic disorders, more than 3,807 diseases (the majority >2,500 being rare) have been characterized at the molecular level, and for over 3,500 of these the genetic causes are still unknown. Some genetic conditions are related to a single gene (e. g. cystic fibrosis), but most genetic diseases are characterized by a great heterogeneity, each of which can be caused by mutations in one of several genes (Intellectual disability > 300 genes, Deafness > 60 genes, Mitochondrial diseases >300 genes). The current molecular diagnosis of clinical phenotypic heterogeneity in genetic conditions is laborious and expensive because it involves the sequential analysis of several genes.

2. The Genetic Diagnosis: A Multi Technical Approach for Neurodevelopmental and Metabolic Disorders

The advance of modern genomics has changed the health economic decisions concerning genetic screening, with costs- per-megabase for DNA sequencing falling at a faster rate than predicted. Although individually rare, Mendelian diseases collectively account for a significant percentage of infant mortality and pediatric hospitalizations [10]. The speed with which genomics is becoming clinically relevant and the increasing power of the new sequencing technologies led to its rapid implementation in clinical settings. Assuming that new technology automatically translates into improved patient care is idealistic.

The ability to amplify DNA by polymerase chain reaction (PCR) has revolutionized our ability to test for genetic mutations. Many different assay systems are based on PCR for analyzing the amplified DNA or RNA. These techniques are sensitive, reliable and can be performed easily on different material: blood, skin or muscle biopsies, tumour, foetus, and even single cells from blastomeres and polar bodies. The aims of adapting new PCR-based strategies in clinical settings are improving the accuracy, speeding up the diagnosis, the time-consuming and costing without decreasing the sensitivity or specificity. Understanding the type of mutations, the incidence of *de novo* mutations or genetic rearrangements is essential to accurately select the best molecular technique for the detection of carrier in pre-natal diagnosis. Techniques involving PCR-based amplification are successfully used for mutation screening in the majority of diseases. It can be combined for the detection of the presence or absence of restriction sites, electrophoretic mobility shift or sizing analysis, as in single strand conformation polymorphism (SSCP) or in denaturing gradient gel electrophoresis (DGGE). Computer-assisted highly sensitive mutation detection is also performed, for the above techniques, by means of fluorescent PCR for allelic specific discrimination. Recent advances in the development of quantitative real-time

PCR (qPCR)-based diagnostic tools allow detection and quantification of gene or exon dosage.

An alternative procedure to mutation-directed protocols for complex genetic mutations is the use of fluorescent multiplex PCR. Indirect diagnosis performed by the use of polymorphic microsatellite markers, allowing identification of the pathogenic haplotype instead of the mutation [11, 12] as for instance in some spinal muscular atrophy or Duchenne muscular dystrophy families. For diseases involving a heterogeneous spectrum of identified mutations, such as cystic fibrosis, autosomal non-syndromic intellectual disability (*SYNGAP1*) or Rett Syndrome (*MECP2*), the development of a mutation-based strategy is not practical and sequencing of the entire coding sequence is recommended to facilitate mutation detection.

Multiplex ligation-dependent probe amplification (MLPA) is a variation of the multiplex polymerase chain reaction that permits multiple targets (exons) to be amplified with only a single primer pair [13]. Specific fluorescent probes consist of two unique oligonucleotides which recognise adjacent target sites on the DNA and each amplicon generates a fluorescent peak which can be detected by a capillary sequencer. Comparing the peak pattern obtained on a given sample with those obtained on various control samples, allow the relative quantity of each target to be determined. This technique is commonly used to detect chromosomal anomalies in cell and tissue samples, [14] detection of gene copy number [13], detection of duplications and deletions in disease-related genes such as *DMD, MECP2, BRCA1, BRCA2,* etc. It has replaced the need to use of southern blot in many diseases [15-17].

Expansion repeat disorders such as of the glutamine codon (CAG) in spinocerebellar ataxias, or the trinucleotide repeat in non-coding regions, as the CGG in Fragile-X syndrome, the GAA in Friedreich ataxia or the CTG in myotonic dystrophy type I, are a special type of mutation. The unstable dynamic nature of those mutations requires the use of a multimodal approach [18]. Molecular tests for the diagnosis and carrier detection often combine PCR for the repeat expansion and triplet repeat primed PCR (TP-PCR) analysis of genomic DNA, as well as southern blot.

The detection rate of the screening test, as well as the frequency of the mutation in the study population and the technical limitations of the procedure, will determine the usefulness of a positive or negative result. Preimplantation testing provides a paradigm for the ease of use of PCR-based testing, yet also underscores the problems encountered with genetic screening because of the multitude of possible mutations and the possible misinterpretation of results [12].

Molecular diagnostic testing is currently available for only a certain number of disorders and with the increasing number of new genes associated to human diseases, it is becoming more and more challenging to provide cost effective molecular testing [10]. Preconception screening, together with genetic counselling of carriers, has resulted in remarkable declines in the incidence of several severe recessive diseases in populations at risk [19, 20]. However, extensive preconception screening and molecular diagnostic testing has been limited to targeted population or family history–directed individual and is still impractical for many disorders [20].

2. 1. Mendelian Single Gene Disorders: At the Mutation or Gene Level

Single-gene disorders have a straight forward inheritance pattern, and the genetic causes can be traced to changes in specific individual genes. A particular disorder may be rare; however, as a group of disease-causing genes, single-gene disorders are responsible for a significant percentage of pediatric diseases [21]. About 1% of the approximately 4 million annual live births in the United States will have a single gene disorder that requires intensive clinical investigation, specific medical treatment and hospitalization [22, 23]. Based on the location of the relevant genes, single-gene traits can be divided into autosomal or sex-linked inheritance. Autosomal inheritance, depending on whether one or two mutant alleles are required to cause the disease phenotype, can be classified as autosomal dominant or autosomal recessive. Each of these single gene disorders, called Mendelian traits or diseases, are relatively uncommon. The frequency often varies with ethnic background, with each ethnic group having one or more Mendelian traits in higher frequency when compared to the other ethnic groups. For example, cystic fibrosis has a frequency of about one in 2,000 births in Americans descended from western European Caucasians [10] but is much rarer in African-Americans descent while sickle cell anemia has a frequency of about 1/600 births in African-American, but is rare in Caucasians [24]. Just to name a few, Mediterranean descent have a high frequency of thalassemia [25]; Eastern European Jews have a high frequency of Tay-Sachs disease [26, 27]; French Canadians from Quebec have a high frequency of tyrosinemia [28], all when compared to other ethnic groups. It has been estimated, regardless of the ethnicity, that each healthy individual is carrying between 1 and 8 mutations which, if found in the homozygous state would result in the expression of a Mendelian recessive disease [10]. Since each human genome has 22,333 genes it is unlikely that any two unrelated individuals would be carrying the same mutations, even if they are from the same ethnic background. This explains why most Mendelian diseases are rare, affecting about 1/10,000 to 1/100,000 live births [21, 29].

In Quebec, the distribution of Mendelian diseases is due to local founder effects caused by the stratification of the contemporary French Canadian gene pool. The migration of a small number of French individuals from France to Quebec created a founder effect. Subsequent inland migrations have created smaller regional founder effects [30, 31]. The limited size of the population favoured genetic drift, and the social context encouraged endogamy, only few unions were reported between French Canadians with English and other immigrants [31]. The French-Canadian population of Quebec, currently about 6 million people, descends from about 7,798 immigrant founders who arrived in Quebec between 1608 and 1759 [31]. Recent studies showed that the Quebec population structure through the analysis of the genetic contribution of the first French settlers can be partitioned in eight regions, and they contributed to over 90% of gene pools in seven out of those eight regions [32]. This particular local genetic effect highlights the importance of considering the geographic origin of samples in the design of genetic testing in Quebec [33]. The conditions under which the peopling of Quebec was made, have favoured changes in the frequency of certain alleles in comparison with the French original population. As a result, certain genetic diseases are specific or more prevalent to the Quebec population. The prevalence and distribution of genetic diseases in Quebec is an essential factor to consider in clinical practice and particularly in differential diagnostic to prioritize molecular investigations [20].

This founder effect has impacted our molecular diagnostic testing system and is still a key factor when developing new diagnostic test for a genetic disease. The prevalence of the disease and the nature of the mutations found in the Quebec population need to be taken into account. The performance of the test depends on how well it accounts for the particularities of the disease in the French Canadians but more specifically depending on the regional founder effect [31]. The current changes in the immigration and the increased admixture is bringing new challenges in the differential diagnosis and amelioration of our molecular genetic testing system.

Over thirty Mendelian diseases have a high prevalence in the Quebec population [31, 34, 35]. Before the advances in sequencing technologies, it was believed that some of the disease were almost exclusive to the French Canadians, as autosomal recessive spastic ataxia of Charlevoix-Saguenay (ARSACS; MIM 270550), hereditary motor and sensory neuropathy with agenesis of the corpus callosum (ACCPN; MIM 218000) or French-Canadian-type Leigh syndrome (MIM 220111) [36-38]. Taking ARSACS as an example, after extensive resequencing of the responsible gene, *SACS*, it became evident that ARSACS was not limited to Quebec, and more than 100 different pathogenic mutations have now been identified worldwide [39, 40]. ARSACS is believed to be underdiagnosed in patients with atypical phenotypes and recent data on exome sequencing suggest a presumably high frequency of allele carriers around the world [39-42].

Admixture has more and more impact on the genetic characteristics of disease in French Canadians. Other ethnic groups have deep roots in the province. Many immigrant communities that are established mainly in and near Montreal now account for over 15% of the Quebec population. First Nations groups in Quebec have specific genetic diseases with three autosomal recessive conditions that have been well documented: Cree leukoencephalopathy [MIM 603896], Cree encephalitis [MIM 608505], North American Indian Childhood Cirrhosis [MIM 604901]. In Cree communities, the carrier frequency of Cree leukoencephalopathy is estimated at 1/10, 1/30 for Cree encephalitis and, 9/100 for the North American Indian Childhood Cirrhosis [43, 44]. However, universal screening is currently offered only in the neonatal period for phenylketonuria, tyrosinaemia type I, medium-chain acyl-coenzyme A dehydrogenase deficiency [MCAD] and congenital hypothyroidism in addition to selected inborn errors of metabolism by urine screening [45, 46]. Effective targeted carrier-screening programs are provided in some specific communities but are often limited to individuals with a family history of recessive diseases. For public health, the genetic structure of Quebec presents major challenge for genetic screening but brings also opportunities for gene identification studies, clinical genetics research and practice. In the next section, we will discuss our experience over time with the implementation of effective targeted genetic testing and carrier-screening programs.

2. 1. 1. Targeted Mutations

The French-Canadian population of Quebec evolved as a mosaic of layered founder effects which has stimulated the development and feasibility of population-based carrier screening for at-risk individuals. For most of the Mendelian diseases in Quebec, the mutant founder alleles are characteristic and often unique. Usually one or two founder mutations account for 90% of

French-Canadian alleles, depending on the region. However, founder mutations panels do exist for some disorders in Quebec such as for familial hypercholesterolemia, hereditary breast cancer, etc.

A good example of a mutant founder allele in Quebec is the Cree encephalitis, a severe early-onset progressive neurological disorder in an inbred Canadian Aboriginal community [MIM 608505]. The symptoms appear within the first few weeks of life and children usually die in infancy or early childhood. The main neurological symptoms are acquired microcephaly, mental retardation, cerebral atrophy with white matter changes, cerebral calcification, and chronic cerebrospinal fluid lymphocytosis. Cree encephalitis shows phenotypic overlap with Aicardi-Goutières syndrome with elevated levels of IFN-a in cerebrospinal fluid [47, 48]. The gene responsible, named *TREX1* (3-prime repair exonuclease 1, MIM 606609), has only one coding exon and is located on 3p21.31. The mutation causing Cree encephalitis is the p. Arg164Ter (c. 490C>T) [49]. The molecular genetic diagnosis consists of determining the presence or absence of p. Arg164Ter in symptomatic patients. CREE encephalitis is among the leading causes of death of Cree infants. Cree carrier rates of this mutation are estimated to be 2-3 individual out of twenty meaning that about one in 300 births will be affected.

In 2006, a genetic screening program was developed and managed by the Cree Health Board to identify carriers of the mutation p. Arg164Ter in the *TREX1* gene. A second autosomal recessive condition is included in the Cree population genetic screening, Cree leukoencephalopathy [MIM 603896]. All patients are homozygous for the mutation p. Glu584Ala in the translation-initiation factor *EIF2B5* gene causing childhood ataxia with central hypomyelination and vanishing white matter disease [CACH/VWM] [50]. Both tests are performed in our molecular diagnostic laboratory at CHU Sainte-Justine. To date, >500 individuals have been tested. Pregnant woman and teenagers in High School are the targeted population for this genetic test. In fact, the rate of teenage pregnancy is very high in Cree population (23% among those aged <20, compared to 4.7% for the rest of Quebec).

Another autosomal recessive disorder caused by a founder mutation in Quebec is the Spastic ataxia of Charlevoix-Saguenay, more commonly known as ARSACS [MIM 270550]. This condition was first observed in people of the Charlevoix-Saguenay region of Quebec, Canada [51]. ARSACS is a debilitating progressive childhood neurological disorder affecting the spinal cord and peripheral nerves. Most patients become wheelchair-bound but cognitive function is usually not affected. The incidence of ARSACS in the Charlevoix-Saguenay region of Quebec is estimated to be 1 in 1,500 to 2,000 individuals. Genetically confirmed cases of ARSACS have now been described in Italy, Spain, Tunisia, France, Belgium, Hungary, Morocco, Turkey, Serbia, other provinces of Canada, Netherlands, United Kingdom, Algeria and Japan. Therefore, ARSACS has a worldwide distribution but its prevalence in most countries is still unknown, except in the Netherlands were it may be as frequent as Friedreich Ataxia [52]. ARSACS is caused by mutations in the gene encoding the SACSIN protein (*SACS*) located on 13q12. 12. Two major mutations were described in the Charlevoix-Saguenay region of Quebec representing 96,3% of the cause of ARSACS. Other types of mutation such as large deletion of 1.54 MB have been reported in some cases [53].

For carrier status genetic testing at the mutation level is the most appropriate method. However, with the growing number of clinically presumed ARSACS cases with a single mutation or no mutations identified, a systematic search for novel *SACS* point mutations and genomic rearrangements for the diagnosis of ARSACS is recommended. Recent reports demonstrated the clinical and genetic heterogeneity among ARSACS patients, now considered to be a common form of spastic ataxia worldwide, the challenge required a multimodal approach to rapidly screen larger numbers of samples and enable detection of both point mutations and deletions. Testing at the mutation level is no longer the optimal methods with the exception of individuals with a clear genealogy to the Charlevoix Saguenay region of Quebec.

2. 1. 2. Targeted Gene

Rett Syndrome (RTT) is one of the most common causes of severe intellectual disability in females, with an incidence of 1:12,500 female live births. Most RTT cases are sporadic with no familial history of the disease. We now know that the majority of RTT patients have a *de novo* mutation in the X-linked gene methyl CpG binding protein 2 (*MECP2*) [54]. Very few families with multiple affected individuals, typical siblings, with Rett Syndrome have been described (for examples [55, 56]) where asymptomatic transmitting mothers have skewed X-chromosome inactivation [56], or parental gonadal mosaicism [57]).

Based on the RettBASE IRSF *MECP2* Variation Database, a database merging mutations and polymorphism data, there are more than 4,000 reported different variants in *MECP2* gene. Given that mutation in any of the 4 exons of *MECP2* can lead to the disease, the targeted gene screening approach is necessary for molecular diagnosis purpose. The same approach is use for autosomal non-syndromic intellectual disability and *SYNGAP1* gene [58].

2. 2. Genetic Disease Panels: The Intermediate Decision

In recent years, there has been increasing recognition of the importance of inherited gene mutations as a cause of neurodevelopmental and metabolic diseases. Substantial progress has been made in elucidating the molecular defects underpinning these diseases, but the impact of these discoveries on clinical management is often unclear and the increasing number of genes associated with these features causes limitations of testing. The clinical spectrum of these diseases is vast and non-specific phenotype presentations are frequent.

In 1975, at the initiative of Dr Charles Scriver, the Province of Quebec implemented an integrated program for the diagnosis, counseling and treatment of hereditary metabolic diseases. Similar programs were implemented worldwide in order for physicians not to miss a treatable disorder. Clinical expression can be acute or systemic or can involve a specific organ, and can strike in the neonatal period or later and intermittently from infancy to late adulthood. During the last half century, many new disorders have been discovered and many therapeutic procedures have been tried. Some have been life-saving; others are still experimental. The long-term outcome of patient with metabolic metabolic patients is variable and requires muti-diciplinary management and treatment [59, 60]. Several factors can impact on the availability and utility of genetic panel testing, such as access to testing, cost of testing and the mutation detection rate. There are over 100 different inborn errors of metabolism, including all the disorders detected by expanded newborn screening for which genetic causes have been identified. Traditionally the inherited metabolic diseases were categorized as disorders of carbohydrate metabolism, amino acid metabolism, organic acid metabolism or lysosomal storage diseases [61]. In recent decades, hundreds of new inherited disorders of metabolism have been discovered and the categories have proliferated. Genetic testing is also indicated in some multisystem disorders with frequent metabolic involvement, however genetic testing is not so straightforward for many of the other conditions. The overall incidence of the inborn errors of metabolism were estimated to be 70 per 100,000 live births or 1 in 1,400 [62]. As the number of inherited metabolic diseases that are included in state-based or province newborn screening programs continues to increase, ensuring the quality and delivery of testing services remain a continuous challenge. The vast majority of inborn errors of metabolism are inherited in an autosomal recessive manner. In the case of diseases carried on the X chromosome, males are usually more severely affected because they have only one copy. Females who have one X chromosome with the gene defect and one without may or may not develop symptoms.

Multi-exon or gene panel testing has proven efficient in cystic fibrosis (mutation panels; focussing on exon 6, 12, 23, 40, 97, 102 etc.) in the form of Ashkenazi disease panels and in the Charlevoix-Saguenay disease panel as described in the previous section. Now, molecular diagnostic laboratories offer panels for Comprehensive Carrier Screen Panels (comprising 80 – 100 conditions, between 350–500 variants). This multi-gene panel testing is becoming more and more advantageous for heterogeneous disorders such as the inborn inherited metabolic disorders. Historically, the development of carrier panels or diagnostic testing panels was subject to the recommendation of professional committees such as the American College of Medical Genetics and Genomics (ACMG) and the College of American Pathologists (CAP). Disease selection for the development of a diagnostic testing panel must take several points into consideration; 1) the disorder is clinically severe, 2) there is a high frequency of carriers in the screened population, 3) the availability of a reliable test with a high specificity and sensitivity, 4) the availability of prenatal diagnosis & access to genetic counselling, 5) genes recommended for screening by ACMG, 6) Autosomal recessive, X-linked diseases or with high de novo mutation rate [63-65].

As another example of multi-gene testing panel, in the past years, several comparative genomic hybridization studies have suggested recurrent rearrangements in synaptic and neurodevelopment genes. Rare CNVs at numerous loci are involved in the cause of mental retardation, autism, and schizophrenia [66-69]. In addition to those, Angelman syndrome and RTT have significant phenotypic overlap, including acquired microcephaly, loss of purposeful movements, developmental delay or intellectual disability and autistic behaviours. Additional features include scoliosis, epilepsy, poor growth or obesity, and irregular breathing. There is also a broad clinical variability in the severity of the disease [70, 71]. *MECP2* mutations are present in 70-90% of females with classic RTT and approximately 20% of females with atypical RTT [72]. Partial deletions of *MECP2* are found in approximately 16% of girls with classic or atypical RTT. *CDKL5* mutations have been demonstrated in a broad spectrum of phenotypes including atypical RTT with infantile spasms [73]. Mutations of the *MEF2C* gene have been identified in patients with severe mental retardation, stereotypic movements, hypotonia, and epilepsy [74]. Abnormalities of the *FOXG1* gene have been identified in patients with the congenital variant of RTT [75]. Patients with the congenital variant of RTT have clinical features similar to classic RTT, but hypotonia and severe developmental delay starts in the first months of life [75]. Phenotypic overlap exists between patients with RTT/Angelman or atypical RTT led to a recommendation for sequencing all coding exons and the intron/exon boundaries of MECP2, CDKL5, MEF2C, and *FOXG1* as well as deletion/duplication analysis for all four genes by array-CGH or MLPA.

In the 1990's, gene testing was mostly done a research-based activity. After the Human Genome Project, companies started offering gene sequencing (Beginning of commercially availability for gene sequencing, Ambry Genetics, City of Hope, GeneDx, Mayo, Baylor, etc.). In the middle of 2000, Sanger sequencing for individual genes or a small group of related genes were offered. Mini-sequencing technologies emerged and let the promise of a more flexible platform to add new mutations or genes to the sequencing chip. However, the cost remains high and to lower the cost, pooling sample became the only feasible way to offer gene testing panel [23]. This reality caused was problematic to small diagnostic laboratories and more importantly for rare orphan diseases.

In the late 2000's, next-generation sequencing started to be used for sequencing medium and large gene panels. These sequencing technologies have benefits (faster diagnosis, reduced costs) and limitations (do not include all genes, incidental findings, variant of unknown significance). In an universal health care system, like in Quebec, diagnostic utility and yields are as important as cost efficiency. Current panels in use are a brief glimpse into the future of molecular diagnostic. Variants of unknown significance will continue to decrease over time and the next generation sequencing will perform better for the detection of copy number and complex rearrangements.

2. 3. The "Omic" Approach, a Comprehensive Interrogation

The introduction of massively parallel sequencing platforms impacted on the cost of sequencing which in turn created, together with progresses in bioinformatics tools, a novel era of study in life sciences. These studies focus on large-scale data and can be divided into three main categories: genomics, proteomics and metabolomics hence the term "omic". Thanks to this technological growth there came new clinical opportunities. These include:

identification of genetic mechanism of heterogeneous disorders (those with dozen and hundreds of genes involved), targeted therapy (especially in cancer), targeted treatment (pharmacogenomics), prenatal screening (example, detection of Trisomy 21 through circulating cell free fetal DNA) and population screening for disease risk.

Genetic variations are a significant determinant of health and response to healthcare. Indeed, 70% of medical decisions rely on clinical laboratory results, and genetic innovations will be a major source of diagnostic and prognostic information in this century. Although the number of genes linked to diseases is continually growing, the number of diagnostic tests has unfortunately lagged behind. Furthermore, complex diseases are expected to be caused by rare variations in a large number of functionally linked genes. For example, non-syndromic intellectual disability (NSID), the most frequent form of intellectual disability representing up to 2/3 of all cases, is caused by mutations in at least 34 genes. Together, these still explain less than 10% of sporadic intellectual disability cases, suggesting that the total number of NSID genes is likely to be above 100. Other well-documented examples of diseases showing high genetic heterogeneity include the spinocerebellar ataxias (35 genes), cardiomyopathies (150 genes), retinitis pigmentosa (78 genes), and deafness (160 genes). Comprehensive gene screening assays for each of these diseases is beyond the capabilities of the majority of diagnostic laboratories. The "omics" era holds the potential to solve the current problem of expanding molecular diagnosis by sequencing at once all the genes in the patient. Exome and genome sequencing are becoming cost-effective in a clinical setting as proven by the multiplication of providers offering clinical exome sequencing in the help for diagnosis (ex: Baylor College of Medicine, Ambry Genetics, GeneDx, UMC St Radboud, etc.).

Establishing an accurate diagnosis is the keystone on which medicine is based. However, an accurate diagnosis can only be accomplished through the use of diagnostic tools that are thoroughly validated for many properties (analytical validity, clinical validity, clinical utility, cost/effectiveness and cost-utility) and supported by the appropriate evidence-based data. A definitive diagnosis greatly facilitates the genetic counseling of families with diseases of early onset, leads to timely intervention, improved outcomes, and provides great reassurance to individuals and their families that they suffer from a known and definable disease.

Sequencing the exome of a patient in a clinical basis is of great value in two obvious situations: 1) finding a new mutation in a known gene underlying an "atypical presentation" (many examples can be found in the FORGE project, *Finding of Rare Disease Genes*: http://www. cpgdsconsortium. com/) and 2) identifying mutations in a novel gene. This could only be possible through a comprehensive approach. Moreover, exome sequencing is a faster and cheaper diagnostic method, it decreases the use of tests and additional consultations, is more accurate and offers personalized treatment. For example, a disease (intellectual disabilities) caused by mutations in ~100 genes will require sequencing of each gene, one at a time, at a cost ~ $ 1,000/gene, and will take several years, for a total cost of over 100,000$. With exome sequencing, screening of these 100 genes now costs ~$ 2,000-$ 7,000 (this cost and analysis turn-around time will decrease rapidly over the next few years). An "omic" approach can also have a positive economic impact. As an example, a 7 year old patient who presented with a neurodegenerative clinical picture characterized by spasticity was referred for exome sequencing. Prior to the sequencing of its entire genes, three pediatric neurologists assessed the child and a plethora of clinical examinations, biochemical tests and medical tests were completed. The three neurologist offered different diagnoses such as spastic paraparesis (> 45 known genes), spinocerebellar ataxia (> 15 known genes) and mitochondrial disease (> 300 known genes). The exploration of these assumptions was simply prohibitive as proposed by the clinical genetic testing targeted 20 genes and would have cost > $ 40,000. Exome sequencing, with a cost of ~$ 7000 rapidly led to the causative mutation/gene.

The major hurdle of exome sequencing is the missing or suboptimal exons coverage mainly due to high GC content. One good example is the aortopathies characterized by aortic dilation, which can lead to life threatening aneurysms and/or dissections. An early diagnosis is critical, since timely initiation of pharmacological treatment can slow dilation and prophylactic surgery can prevent aortic dissection or rupture. Mortality rate is 20% for aortic dissections. Mutations in twelve different genes (for a total of 360 exons) are known to be responsible for these diseases. The commercial kit of exome capture SureSelect from Agilent is missing (or suboptimal) 11% exons of these genes. Fortunately, these weaknesses will be overcome with promised technological update.

3. The future of Molecular Diagnostic Lab: What Will Be Kept and What Will Bot

Routine DNA-based diagnostic sequencing for neurodevelopmental or metabolic disorders currently targets well-defined genes or sets of genes that address very specific clinical questions. Today, available services vary greatly depending on the diagnostic laboratory, including Sanger sequencing of complete genes harboring mutations with known, well-described clinical phenotypes or multi-gene testing panels or next-generation sequencing-based disease/phenotype panels. Test ordering depends on the ability of a physician to apply a differential diagnosis; it follows a paradigm where a patient with phenotype A will have a mutation in a known gene causing phenotype A, which will be sequenced. Today's reality is that several genes are known to cause almost any particular phenotype, so gene panels can be used to sequence multiple genes simultaneously, or in a linear approach where the most commonly mutated genes are sequenced first, and if a negative result is obtained, the remainder are sequenced. The recent advances of next-generation sequencing enable the laboratory to simultaneously sequence a large number of genes at a significantly decreasing cost per gene. This raises multiple questions - is there a point at which additional sequencing diminishes the clinical utility of the resulting data? Is the proliferation of DNA testing panels enabled by next-generation sequencing beneficial for clinical diagnostics? What will be kept and what will be dropped?

This type of linear strategy is generally very satisfying for the diagnosing clinician as each test in specific population, for example in micro founder effect of Cree subpopulation or Saguenay, has a known value for diagnostic purpose. It also utilizes a focused approach, which tends to prioritize the most common gene involved in a certain phenotype. The genetics field is still young, but the actual genes and associated casual mutations are often incrementally discovered. Newly discovered genes are reported weekly, but it is generally difficult to systematically add them to a laboratory test menu. The diagnostic value from a single, well-characterized gene is relatively high since it is easier to interpret and it minimizes the likelihood that a variant of unknown significance will be discovered [64]. Howeer, if a gene is reported only for a few cases, the diagnostic test may not be offered. Offering genetic testing for rare orphan disease has a high relative cost of test development and a low diagnostic yield [41, 63, 64].

As described in the previous section, some specific mutation panels will remains for at-risk population until the cost of next-generation sequencing decreases further. In addition, triplet-repeat diseases because of the unstable dynamic nature of those mutations will continue to require the use of a multimodal approach.

Next-generation sequencing is really moving the field of towards comprehensive gene panels. Expanded comprehensive diagnostic gene panels have several advantages. First of all, the design of an expanded diagnostic panel is quite straightforward and begins by identifying all of the genes involved in the disease(s) being targeted. Adding genes to the list is simple and does not requir additional cost, unless enrichment is needed [63]. More comprehensive gene panel tests will simplify test ordering by consolidating all candidate genes into a single diagnostic test. By taking a more comprehensive approach, the sensitivity of the test increases and the rate of molecular under diagnosis decreases by including genes with low-frequency. It will also allow the identificaiton of in the causative gene in patients with an atypical phenotypic presentation. Moreover, with the advent of next-generation sequencing, an expanded neurodevelopmental and metabolic molecular diagnostic sequencing panel becomes economically feasible, allowing diagnosis of extremely rare orphan disease in one genetic test.

Next-generation sequencing introduces some interpretative challenges which are not new to diagnostic testing, but the scale of incidental findings or number of variants of unknown significance will be far greater than any previous test. Sanger sequencing typically identifies three categories of mutations: known non-pathogenic or benign polymorphisms; known pathogenic mutations; and variants of unknown significance. Variants identified as known pathogenic mutations are straightforward to interpret as in most cases, there are clearly indicated for clinical action [76]. With time, and better variant databases, the experience gained using next-generation sequencing testing panels will provide more data to improve variant interpretation and, over time, reduce the total number of variants of unknown significance encountered during next-generation sequencing diagnostic test [76]. The molecular clinical approach should focus on immediate clinical benefit of next-generation sequencing, such as the ability to offer comprehensive gene panels, to provide a first-pass diagnostic yield, and to limit gene list to genes of known diagnostic value [64]. This will aid the interpretation of the data set in a way that the healthcare provider will be able to understand and utilize.

4. The Non-Targeted Genetic Strategy Generates New Ethical Challenges

Particular emphasis has been placed in recent years on the identification of rare disease genes due to the availability of new genomic sequencing technologies. As seen earlier, these technologies greatly facilitate the search for causative disease mutations since they allow the analysis of the 22,333 genes at once in a short period of time for a reasonable cost. As a result, more genetic mechanisms involved in rare diseases are known, and a portion of these findings are already translated into diagnostic tools. Indeed, the clinical sequencing (referring to exome or genome sequencing) is rapidly being integrated into the practice of medicine. Although technically feasible and proven to be successful [77, 78], several challenges remain surrounding the use of clinical sequencing. Nonetheless molecular diagnosis using genomic and genetic methods such as microarray CGH, exome and genome sequencing have in common several ethical, legal and social concerns with other methods of medical investigation. Because of the unprecedentedly large amount of information generated by these comprehensive tests on an individual, the concerns (e. g. content of the consent form), incidental findings, returning results to the patient, etc. are amplified.

4. 1. Consent Form

Consent forms need to be adapted to the new generation of sequencing clinical test. The main addition to the current medical consent form is the extent to which the patients would have control/access to the whole data (exome/genome), what results might be returned to the patient and what are the potential risks. The consent form should also inform the patient about the possible risk of discovering unwanted findings (unrelated to the original medical investigation). For instance, the exome analysis may reveal that the patient is likely to get a serious and untreatable disease.

4. 2. Data Storage and Sharing

Next-generation sequencing technologies generate terabytes of data, and storing this amount of data will constitute a challenge in itself but also raised ethical issues. The genetic information of individual genome that is derived from the clinical sequencing is an accessible and robust "login" into a patient identity. It can give information on the patient's relatives, ethnic groups, and more. The Wellcome Trust and the National Institute of Health, two respective Institutions whose mission is to operate open access genetic and genomic databases for the benefits of researchers and patients, now restricted and controlled their web database access. They even formed specific committees to oversee the circulation of the data included in their databases. This was effective following the conclusion from the study of Homer and colleagues in 2008 demonstrating that genome-wide association derived data is sufficient to re-identify an individual that have participated in a study [79]. They concluded that anonymizing data was unsatisfactory to protect the confidentiality of research participants. Because the understanding, diagnosed and treatment of complex diseases such as cancer or pediatric disorders required comparison of genetic information from several hundreds and often thousands of individuals, genetic data sharing is critical for research and molecular diagnosis. To further illustrate the significance of protecting information derived from whole genome sequencing of individual, the Presidential Commission for the Study of Bioethical Issues published a 150-page report, on the Privacy and progress in the era of whole genome sequencing. They concluded *"to realize the enormous promise that whole genome sequencing holds for advancing clinical care and the greater public good, individual interests in privacy must be respected and secured. As the scientific community works to bring the cost of whole genome sequencing down from millions per test to less than the cost of many standard diagnostic tests today, the Commission recognizes that whole genome sequencing and its increased use in research and the clinic could yield major advances in health care. However it could also raise ethical dilemmas. The Commission offers a dozen timely proactive recommendations that will help craft policies that are flexible enough to ensure progress and responsive enough to protect privacy. "* Finally, researchers and policy-makers need to find methods to protect individuals' genomic data while still being able to share information.

4. 3. Incidental Findings

As mentioned in earlier sections, clinical exome and genome sequencing present several advantages for clinical practice and for the patient. The main advantage discussed is the nondiscrimination on the selection of candidate's genes that will be analyzed. However, these methods can yield incidental medical information not related to the principal medical target. For instance, the genome of a child investigated for a rare form of deafness can lead to the discovery of a known mutation leading to a late onset neurodegenerative disorder. In addition, the patient can be found to be carrier of a recessive lethal disease. These "secondary" findings have been the subject of many discussions and recently the American College of Medical Genetics and Genomics whose mission is to improve health through medical genetics released it's highly anticipated "Recommendations on Incidental Findings in Clinical Exome and Genome Sequencing".

In addition to fortuitous findings, whole genome sequencing raises great social concerns that still need to be resolved, such as possible forms of discrimination. Efforts have been made to overcome this phenomenon since the Genetic Information Nondiscrimination Act was signed in the USA in 2008 to protect individuals from improper use of genetics information with regard to health insurance and employment [69]. Emerging discoveries of susceptibility genes and gene variants associated with major neurological or psychiatric disorders are likely to challenge the existing ethical guidelines. It is up to researchers, health professionals and experts in the Ethical, Economic, Environmental, Legal, and Social aspects of genomics [GE3LS] to manage this growing scientific knowledge in a way that prioritizes protection of research participants and improves patient care.

Conclusion and Future

The successful use of whole-genome and exome sequencing for diagnosis has been amply confirmed by numerous studies. Whether targeted gene, gene panel approaches and exome sequencing will be entirely replaced by whole-genome sequencing is still unknown. Compared to traditional methods, it is now well accepted that exome sequencing has fewer false positives, and a greater sensitivity due to the higher coverage achieved when focusing only on a small fraction of the genome. Exome and whole-genome sequencing allow the discovery of point mutations and small deletion. With advanced bioinformatics tool it is now feasible to detect larger insertions and deletions (CNVs), currently detected using the microarrays technologies. We think that these "cytogenetic" methods, including karyotyping and molecular diagnosis, will merge. We also think that the current modalities of next-generation sequencing will eventually be replaced by genome sequencing. Recently,

Saunders and colleague showed the feasibility of using whole-genome sequencing in neonatal intensive care units to screen for genetic disease diagnosis [77]. They described a 50-hour delay in the diagnosis of genetic disorders using whole genome sequencing.

Some challenges will subsist: mosaicism, balanced translocation and complex diseases (unknown mode of transmission). However, an important challenge in using whole-genome sequencing will be the interpretation of the data linking a genetic variation to the disease/phenotype and all ethical issues around it.

The rapid development of sequencing is now positively impacting prenatal diagnosis. Since the discovery of cell-free fetal nucleic acids circulating in the blood of pregnant women, [80] combined with next-generation DNA sequencing technologies, it is now possible to do early, noninvasive prenatal genetic testing [81]. Clinical applications of these methods already include fetal sex determination and blood group typing [82]. Ongoing research is currently evaluating the use of this approach for noninvasive detection of trisomies [83]. Other uses being explored are the detection of single-gene disorders, chromosomal abnormalities and inheritance of parental polymorphisms across the whole fetal genome.

Use of preimplantation genetic diagnosis and preimplantation genetic screening using next-generation sequencing can provide blastocyst preimplantation genetic diagnosis (PGD) results with high level of consistency with established diagnostic methods. Furthermore, single-gene disorder screening by next-generation sequencing could be performed in parallel with qPCR-based comprehensive chromosome screening or array SNP-CGH. Several studies showed that next-generation sequencing could serve as an essential PGD tools for further development of this important and emerging field [84]. Next-generation sequencing data provides a unique opportunity to evaluate multiple genomic loci and multiple samples on one experiment (e. g foetus can be tested in parallel with the parents). Next-generation sequencing might also be useful for simultaneous evaluation of aneuploidy, single-gene disorders, and translocations from the same biopsy without the need for multiple technological platforms [85]. Clearly, the fertility field has seen intensive efforts to significantly improve and validate better state-of-the-art in vitro fecundation techniques. Martin J. et al. suggested that IVF techniques coupled with deep comprehensive diagnosis/screening methods using NGS should result in high implantation and live birth rates [85]. Nevertheless, there is always room for improvements; it is important to be prudent, recognize the limits of sequence depth necessary to maintain accuracy, and variation in sequencing depth across different genomic loci that is critical to its clinical application in PGD. We expect that whole-genome sequencing will allow the identification of genetic variants that will determine an individual's risk for developing diseases, including neurological or psychiatric disorders. In fact, the continuing reduction in sequencing costs may lead to replacement of most of the other currently used approaches.

Acknowledgements

We are very grateful for the support of our Institution. We wish to thank our colleagues, Mélanie Lafleur, Françoise Couture, Sylvie Filiatrault, Josée Gauthier, Janique Ladouceur, Caroline Deschaînes, Marie-Josée Lassonde, Julie Ménard, Li Fan and Marie-Lyne Darveau for their technical and administrative work in the Molecular Diagnostic Lab. We also acknowledge the work/input of our collaborative clinicians: Drs. Jacques Michaud, Grant Mitchell, Jean-François Soucy, Dorothée Dal Soglio, Luc Oligny, George-Etienne Rivard and Sonia Cellot for their help in the discovery, development and for working in close collaboration to improve the use of genetics and genomic knowledge and tools in today's medicine.

References

[1]   Sanger, F. , S. Nicklen, and A. R. Coulson, DNA sequencing with chain-terminating inhibitors. *Proc Natl Acad Sci* U S A, 1977. *74*(12): p. 5463-7.

[2]   Pruitt, K. D. , et al. , NCBI Reference Sequences: current status, policy and new initiatives. *Nucleic Acids Res,* 2009. *37*(Database issue): p. D32-6.

[3]   *The Human Genome Project.* http://genomics. energy. gov. , 1999.

[4]   Awadalla, P. , et al. , Direct measure of the de novo mutation rate in autism and schizophrenia cohorts. *Am J Hum Genet,* 2010. *87*(3): p. 316-24.

[5]   Hamdan, F. F. , et al. , Excess of de novo deleterious mutations in genes associated with glutamatergic systems in nonsyndromic intellectual disability. *Am J Hum Genet,* 2011. *88*(3): p. 306-16.

[6]   Griswold, A. J. , et al. , Evaluation of Copy Number Variations Reveals Novel Candidate Genes in Autism Spectrum Disorder Associated Pathways. *Hum Mol Genet,* 2012.

[7]   Elia, J. , et al. , Genome-wide copy number variation study associates metabotropic glutamate receptor gene networks with attention deficit hyperactivity disorder. *Nat Genet,* 2012. *44*(1): p. 78-84.

[8]   Pfundt, R. and J. A. Veltman, Structural genomic variation in intellectual disability. *Methods Mol Biol,* 2012. *838*: p. 77-95.

[9]   Malhotra, D. , et al. , High frequencies of de novo CNVs in bipolar disorder and schizophrenia. *Neuron,* 2011. *72*(6): p. 951-63.

[10]   Kingsmore, S. , Comprehensive carrier screening and molecular diagnostic testing for recessive childhood diseases. *PLoS Curr,* 2012: p. e4f9877ab8ffa9.

[11]   Dreesen, J. C. , et al. , Multiplex PCR of polymorphic markers flanking the CFTR gene; a general approach for preimplantation genetic diagnosis of cystic fibrosis. *Mol Hum Reprod,* 2000. *6*(5): p. 391-6.

[12]   Moutou, C. , N. Gardes, and S. Viville, Multiplex PCR combining deltaF508 mutation and intragenic microsatellites of the CFTR gene for pre-implantation genetic diagnosis (PGD) of cystic fibrosis. *Eur J Hum Genet,* 2002. *10*(4): p. 231-8.

[13]   Schouten, J. P. , et al. , Relative quantification of 40 nucleic acid sequences by multiplex ligation-dependent probe amplification. *Nucleic Acids Res,* 2002. *30*(12): p. e57.

[14]   Hochstenbach, R. , et al. , Rapid detection of chromosomal aneuploidies in uncultured amniocytes by multiplex ligation-dependent probe amplification (MLPA). *Prenat Diagn,* 2005. *25*(11): p. 1032-9.

[15]   Marzese, D. M. , et al. , Detection of deletions and duplications in the Duchenne muscular dystrophy gene by the molecular method MLPA in the first Argentine affected families. *Genet Mol Res,* 2008. *7*(1): p. 223-33.

[16]   Engert, S. , et al. , MLPA screening in the BRCA1 gene from 1,506 German hereditary breast cancer cases: novel deletions, frequent involvement of exon 17, and occurrence in single early-onset cases. *Hum Mutat,* 2008. *29*(7): p. 948-58.

[17]   Makrythanasis, P. , et al. , De novo duplication of MECP2 in a girl with mental retardation and no obvious dysmorphic features. *Clin Genet,* 2010. *78*(2): p. 175-80.

[18]   Thompson, T. , *Genetics in Medicine.* 7th ed, ed. S. Elsevier. 2007, Philadelphia: Elsevier. 567.

[19]   Andermann, A. , I. Blancquaert, and V. Dery, Genetic screening: a conceptual framework for programmes and policy-making. J *Health Serv Res Policy*, 2010. *15*(2): p. 90-7.

[20]   Laberge, A. M. , et al. , Population history and its impact on medical genetics in Quebec. *Clin Genet*, 2005. *68*(4): p. 287-301.

[21]   Rodelsperger, C. , et al. , Identity-by-descent filtering of exome sequence data for disease-gene identification in autosomal recessive disorders. *Bioinformatics*, 2011. *27*(6): p. 829-36.

[22]   Yu, J. H. , et al. , Self-guided management of exome and whole-genome sequencing results: changing the results return model. *Genet Med*, 2013.

[23]   Jamal, S. M. , et al. , Practices and policies of clinical exome sequencing providers: analysis and implications. *Am J Med Genet A*, 2013. *161A*(5): p. 935-50.

[24]   Section on Hematology/Oncology Committee on, G. and P. American Academy of, Health supervision for children with sickle cell disease. *Pediatrics*, 2002. *109*(3): p. 526-35.

[25]   Atanasovska, B. , et al. , Efficient detection of Mediterranean beta-thalassemia mutations by multiplex single-nucleotide primer extension. *PLoS One*, 2012. *7*(10): p. e48167.

[26]   Natowicz, M. R. and E. M. Prence, Heterozygote screening for Tay-Sachs disease: past successes and future challenges. *Curr Opin Pediatr*, 1996. *8*(6): p. 625-9.

[27]   Rozenberg, R. and V. Pereira Lda, The frequency of Tay-Sachs disease causing mutations in the Brazilian Jewish population justifies a carrier screening program. *Sao Paulo Med J*, 2001. *119*(4): p. 146-9.

[28]   Bergeron, P. , C. Laberge, and A. Grenier, Hereditary tyrosinemia in the province of Quebec: prevalence at birth and geographic distribution. *Clin Genet*, 1974. *5*(2): p. 157-62.

[29]   Guillevin, L. , [The national plan for orphan rare diseases: nearly 10 years on]. *Rev Neurol (Paris)*, 2013. *169 Suppl 1*: p. S9-11.

[30]   Bouchard, G. , C. Laberge, and C. R. Scriver, [Demographic reproduction and genetic transmission in the north-east of the province of Quebec (18th-20th centuries)]. *Eur J Popul*, 1988. *4*(1): p. 39-67.

[31]   Laberge, A. M. , [Prevalence and distribution of genetic diseases in Quebec: impact of the past on the present]. *Med Sci (Paris)*, 2007. *23*(11): p. 997-1001.

[32]   Roy-Gagnon, M. H. , et al. , Genomic and genealogical investigation of the French Canadian founder population structure. *Hum Genet*, 2011. *129*(5): p. 521-31.

[33]   Bherer, C. , et al. , Admixed ancestry and stratification of Quebec regional populations. *Am J Phys Anthropol,* 2011. *144*(3): p. 432-41.

[34]   Scriver, C. R. , et al. , Feasibility of chemical screening of urine for neuroblastoma case finding in infancy in Quebec. *CMAJ,* 1987. *136*(9): p. 952-6.

[35]   Scriver, C. R. , Human genetics: lessons from Quebec populations. *Annu Rev Genomics Hum Genet,* 2001. *2*: p. 69-101.

[36]   Engert, J. C. , et al. , ARSACS, a spastic ataxia common in northeastern Quebec, is caused by mutations in a new gene encoding an 11. 5-kb ORF. *Nat Genet,* 2000. *24*(2): p. 120-5.

[37]   Debray, F. G. , et al. , LRPPRC mutations cause a phenotypically distinct form of Leigh syndrome with cytochrome c oxidase deficiency. *J Med Genet,* 2011. *48*(3): p. 183-9.

[38]   Howard, H. C. , et al. , The K-Cl cotransporter KCC3 is mutant in a severe peripheral neuropathy associated with agenesis of the corpus callosum. *Nat Genet,* 2002. *32*(3): p. 384-92.

[39]   Thiffault, I. , et al. , Diversity of ARSACS mutations in French-Canadians. *Can J Neurol Sci,* 2013. *40*(1): p. 61-6.

[40]   Synofzik, M. , et al. , Autosomal recessive spastic ataxia of Charlevoix Saguenay (ARSACS): expanding the genetic, clinical and imaging spectrum. *Orphanet J Rare Dis,* 2013. *8*: p. 41.

[41]   Chen, Z. , et al. , Using next-generation sequencing as a genetic diagnostic tool in rare autosomal recessive neurologic Mendelian disorders. *Neurobiol Aging,* 2013.

[42]   Tzoulis C, J. S. , Haukanes BI, Boman H, Knappskog PM, et al. , Novel SACS Mutations Identified by Whole Exome Sequencing in a Norwegian Family with Autosomal Recessive Spastic Ataxia of Charlevoix-Saguenay. *PLoS One,* 2013. *8*(6): p. e66145.

[43]   Black, D. N. , et al. , Leukoencephalopathy among native Indian infants in northern Quebec and Manitoba. *Ann Neurol,* 1988. *24*(4): p. 490-6.

[44]   Drouin, E. , et al. , North American Indian cirrhosis in children: a review of 30 cases. *J Pediatr Gastroenterol Nutr,* 2000. *31*(4): p. 395-404.

[45]   Laberge, C. , et al. , [The genetic medicine network in Quebec: An integrated program for diagnosis, counseling and treatment of hereditary metabolic diseases]. *Union Med Can,* 1975. *104*(3): p. 428-32.

[46]   Grenier, A. and C. Laberge, [Detection of hereditary metabolic diseases in Quebec]. *Union Med Can,* 1974. *103*(3): p. 453-6.

[47]   Crow, Y. J. , et al. , Cree encephalitis is allelic with Aicardi-Goutieres syndrome: implications for the pathogenesis of disorders of interferon alpha metabolism. *J Med Genet,* 2003. *40*(3): p. 183-7.

[48]   Black, D. N. , et al. , Encephalitis among Cree children in northern Quebec. *Ann Neurol,* 1988. *24*(4): p. 483-9.

[49]   Crow, Y. J. , et al. , Mutations in the gene encoding the 3'-5' DNA exonuclease TREX1 cause Aicardi-Goutieres syndrome at the AGS1 locus. *Nat Genet,* 2006. *38*(8): p. 917-20.

[50]   Fogli, A. , et al. , Cree leukoencephalopathy and CACH/VWM disease are allelic at the EIF2B5 locus. *Ann Neurol,* 2002. *52*(4): p. 506-10.

[51]   Bouchard JP, R. A. , Melancon SB, Mathieu J, Michaud J. , Autosomal recessive spastic ataxia (Charlevoix– Saguenay). *Handbook of ataxia disorders,* Klockgether T, New York: Marcel Dekker, 2000: 311–324. , 2000.

[52]   Vermeer, S. , et al. , ARSACS in the Dutch population: a frequent cause of early-onset cerebellar ataxia. *Neurogenetics,* 2008. *9*(3): p. 207-14.

[53]   Breckpot, J. , et al. , A novel genomic disorder: a deletion of the SACS gene leading to spastic ataxia of Charlevoix-Saguenay. *Eur J Hum Genet,* 2008. *16*(9): p. 1050-4.

[54]   Amir, R. E. , et al. , Rett syndrome is caused by mutations in X-linked MECP2, encoding methyl-CpG-binding protein 2. *Nat Genet,* 1999. *23*(2): p. 185-8.

[55]   Curtis, A. R. , et al. , X chromosome linkage studies in familial Rett syndrome. *Hum Genet,* 1993. *90*(5): p. 551-5.

[56]   Schanen, N. C. , et al. , A new Rett syndrome family consistent with X-linked inheritance expands the X chromosome exclusion map. *Am J Hum Genet,* 1997. *61*(3): p. 634-41.

[57]   Wan, M. , et al. , Rett syndrome and beyond: recurrent spontaneous and familial MECP2 mutations at CpG hotspots. *Am J Hum Genet,* 1999. *65*(6): p. 1520-9.

[58]   Hamdan, F. F. , et al. , Mutations in SYNGAP1 in autosomal nonsyndromic mental retardation. *N Engl J Med,* 2009. *360*(6): p. 599-605.

[59]   Saudubray, J. M. , F. Sedel, and J. H. Walter, Clinical approach to treatable inborn metabolic diseases: an introduction. *J Inherit Metab Dis,* 2006. *29*(2-3): p. 261-74.

[60]   Saudubray, J. M. , Neurometabolic disorders. *J Inherit Metab Dis,* 2009. *32*(5): p. 595-6.

[61]   Fernandes, J. S. , J. M. ; van den Berghe, G. ; Walter, J. H. , *Inborn Metabolic Diseases : Diagnosis and Treatments*. 4th ed, ed. Springer. 2006: Springer.

[62]   Applegarth, D. A. , J. R. Toone, and R. B. Lowry, Incidence of inborn errors of metabolism in British Columbia, 1969-1996. Pediatrics, 2000. *105*(1): p. e10.

[63]   Ku, C. S. , et al. , Exome sequencing: dual role as a discovery and diagnostic tool. *Ann Neurol*, 2012. *71*(1): p. 5-14.

[64]   Klee, E. W. , N. L. Hoppman-Chaney, and M. J. Ferber, Expanding DNA diagnostic panel testing: is more better? *Expert Rev Mol Diagn*, 2011. *11*(7): p. 703-9.

[65]   Levenson, D. , The tricky matter of secondary genomic findings: ACMG plans to issue recommendations. *Am J Med Genet A*, 2012. *158A*(7): p. ix-x.

[66]   Lee, J. A. and J. R. Lupski, Genomic rearrangements and gene copy-number alterations as a cause of nervous system disorders. *Neuron*, 2006. *52*(1): p. 103-21.

[67]   Kusenda, M. and J. Sebat, The role of rare structural variants in the genetics of autism spectrum disorders. *Cytogenet Genome Res*, 2008. *123*(1-4): p. 36-43.

[68]   Guilmatre, A. , et al. , Recurrent rearrangements in synaptic and neurodevelopmental genes and shared biologic pathways in schizophrenia, autism, and mental retardation. *Arch Gen Psychiatry*, 2009. *66*(9): p. 947-56.

[69]   Cantor, R. M. and D. H. Geschwind, Schizophrenia: genome, interrupted. *Neuron*, 2008. *58*(2): p. 165-7.

[70]   Williams, C. A. , et al. , Angelman syndrome: mimicking conditions and phenotypes. *Am J Med Genet*, 2001. *101*(1): p. 59-64.

[71]   Marschik, P. B. , et al. , Changing the perspective on early development of Rett syndrome. *Res Dev Disabil*, 2013. *34*(4): p. 1236-9.

[72]   Neul, J. L. , The relationship of Rett syndrome and MECP2 disorders to autism. Dialogues *Clin Neurosci*, 2012. *14*(3): p. 253-62.

[73]   Maortua, H. , et al. , CDKL5 gene status in female patients with epilepsy and Rett-like features: two new mutations in the catalytic domain. *BMC Med Genet*, 2012. *13*: p. 68.

[74]   Novara, F. , et al. , MEF2C deletions and mutations versus duplications: A clinical comparison. *Eur J Med Genet*, 2013. *56*(5): p. 260-5.

[75] Guerrini, R. and E. Parrini, Epilepsy in Rett syndrome, and CDKL5-
      and FOXG1-gene-related encephalopathies. *Epilepsia*, 2012. *53*(12): p.
      2067-78.
[76] Richards, C. S. , et al. , ACMG recommendations for standards for
      interpretation and reporting of sequence variations: Revisions 2007.
      Genet Med, 2008. *10*(4): p. 294-300.
[77] Saunders, C. J. , et al. , Rapid whole-genome sequencing for genetic
      disease diagnosis in neonatal intensive care units. *Sci Transl Med*, 2012.
      4(154): p. 154ra135.
[78] de Ligt, J. , et al. , Diagnostic exome sequencing in persons with severe
      intellectual disability. *N Engl J Med*, 2012. *367*(20): p. 1921-9.
[79] Homer, N. , et al. , Resolving individuals contributing trace amounts of
      DNA to highly complex mixtures using high-density SNP genotyping
      microarrays. *PLoS Genet*, 2008. *4*(8): p. e1000167.
[80] Lo, Y. M. , et al. , Presence of fetal DNA in maternal plasma and serum.
      Lancet, 1997. *350*(9076): p. 485-7.
[81] Sayres, L. C. and M. K. Cho, Cell-free fetal nucleic acid testing: a
      review of the technology and its applications. *Obstet Gynecol Surv*,
      2011. *66*(7): p. 431-42.
[82] Lo, Y. M. , et al. , Prenatal diagnosis of fetal RhD status by molecular
      analysis of maternal plasma. *N Engl J Med*, 1998. *339*(24): p. 1734-8.
[83] Chiu, R. W. , et al. , Non-invasive prenatal assessment of trisomy 21 by
      multiplexed maternal plasma DNA sequencing: large scale validity
      study. *BMJ*, 2011. *342*: p. c7401.
[84] Treff, N. R. , et al. , Evaluation of targeted next-generation sequencing-
      based preimplantation genetic diagnosis of monogenic disease. *Fertil
      Steril*, 2013. *99*(5): p. 1377-1384 e6.
[85] Martin, J. , et al. , The impact of next-generation sequencing technology
      on preimplantation genetic diagnosis and screening. *Fertil Steril*, 2013.
      99(4): p. 1054-61 e3.

Index

A

access, 154, 155, 161, 162
acetic acid, 13
acetylation, 53, 54, 58, 66, 128
acid, 101, 102, 116, 128, 154
acrocentric chromosome, 14
acute leukemia, 70, 103
acute lymphoblastic leukemia, 105
acute myeloid leukemia, 33, 70, 105, 118
AD, 114, 117, 130
adaptation, 56, 127
adenine, 99
adenocarcinoma, 71, 124, 136, 137
adenoma, 110
adhesion, 54, 75, 100
adipose, 6
adulthood, 7, 154
adults, 66, 138
advancement, 15
aetiology, 44, 99, 131
African-American, 149
aggressiveness, 77
alanine, 29
Algeria, 153
algorithm, 36
allele, 24, 69, 151, 152
alpha-fetoprotein, 138
amino, 58, 145, 154
amino acid, 145, 154

amniotic fluid, 13, 80, 107
androgen, 114
aneuploid, 3, 63, 65
aneuploidy, 16, 21, 48, 54, 56, 107, 127, 164
angiogenesis, 3, 60
annealing, 32
antibody, 27, 39, 64, 68, 76, 78
antigen, 26, 62, 64, 74, 75, 76, 100, 102, 104, 138, 140
antisense, 59
APC, 54, 65, 92, 93, 94, 96, 99
APL, 82
apoptosis, 51, 54, 56, 58, 59, 60, 135
apoptotic pathways, 56
arginine, 2
arrest, 58, 59
Asia, 137
aspirate, 92, 93
aspiration, 10, 16, 93
assessment, 63, 67, 105, 171
astrocytoma, 6
asymptomatic, 75, 153
ataxia, 148, 151, 152, 153, 158, 168, 169
ATP, 30
atrophy, 148, 152
attachment, 21, 55
autism, 145, 146, 155, 165, 170
Autoantibodies, 140
automation, 12, 40, 41

autosomal dominant, 149
autosomal recessive, 149, 151, 152, 155, 167, 168
avian, 101

B

BAC, 20, 23, 35, 36, 120
bacteria, 25
bacterial artificial chromosome, 105
base, 2, 16, 20, 22, 32, 34, 37, 40, 41, 43, 47, 55, 119, 144
base pair, 2, 20, 22, 40, 41, 144
beams, 12
Belgium, 153
benefits, 156, 162
benign, 47, 57, 138, 160
BI, 168
bias, 17
bile duct, 76
biliary tract, 70, 134
bioinformatics, viii, 143, 156, 163
biological fluids, 39
biomarker(s), vii, viii, 1, 2, 5, 38, 39, 44, 57, 60, 61, 62, 64, 65, 66, 67, 69, 71, 72, 73, 74, 77, 78, 79, 83, 97, 98, 99, 103, 118, 120, 130, 131, 133, 135, 136, 137, 139, 140
biopsy, viii, 2, 10, 11, 38, 71, 77, 79, 80, 83, 98, 164
biotechnology, 28
bipolar disorder, 166
birth rate, 164
births, 149, 152, 153, 154
bladder cancer, 75
blood, 6, 7, 10, 11, 12, 13, 67, 69, 72, 74, 76, 79, 80, 96, 98, 135, 147, 164
blood group, 164
blood stream, 72, 74, 98
blood vessels, 6, 12
body fluid, 10, 38, 68, 79, 98
bonds, 59
bone(s), 6, 10, 12, 13, 15, 52, 56, 79, 80
bone marrow, 6, 10, 13, 15, 79, 80
brain, 5, 6, 7, 51

brain cancer, 5
breast cancer, 5, 35, 58, 59, 66, 69, 71, 72, 73, 75, 105, 108, 109, 112, 113, 120, 128, 130, 131, 133, 136, 137, 138, 140, 152, 166
breast carcinoma, 39
breast lumps, 12
breathing, 155

C

calcification, 152
calcium, 78
cancer cells, 3, 10, 26, 39, 47, 53, 54, 57, 60, 66, 68, 74, 98, 116, 122, 127, 134
cancer progression, 73, 123
cancer screening, 136
candidates, 72, 98
CAP, 155
capillary, 29, 31, 38, 117, 145, 148
carbohydrate, 154
carbohydrate metabolism, 154
carcinoembryonic antigen, 80, 139
carcinogenesis, 56, 58, 71, 77, 78, 129
carcinoma, 6, 7, 8, 26, 39, 70, 73, 74, 75, 76, 101, 105, 106, 109, 110, 112, 113, 115, 116, 120, 124, 132, 135, 136, 137, 139, 140
cartilage, 6
castration, 136
CAT scan, 12
Caucasians, 149
causation, 130
CBP, 54, 81, 99
CCND2, 93, 100
cDNA, 20, 24, 93, 111, 121, 144
cell culture, 13, 21, 44
cell cycle, 4, 55, 56, 58, 62, 122, 135
cell death, 4, 51, 53, 58, 132
cell division, 4, 13, 47, 55, 57
cell fate, 128
cell line(s), 21, 26, 35, 44, 50, 52, 54, 59, 69, 112, 113, 115, 123, 124, 125, 129, 133, 134
centromere, 3, 14, 63, 107

centrosome, 48, 56, 127
cerebrospinal fluid, 152
cervical cancer, 10, 51, 113, 124, 137
cervix, 10
challenges, viii, 23, 79, 128, 144, 150, 160,
    161, 164, 167
chemical, 31, 50, 168
chemiluminescence, 30
chemotherapy, 57, 71, 75, 77, 134
childhood, 7, 103, 146, 152, 166
children, 66, 146, 152, 167, 168, 169
cholangiocarcinoma, 133, 134, 135
cholecystitis, 76
cholelithiasis, 76
chondrosarcoma, 6
choriocarcinoma, 76
chorionic villi, 13, 17, 79, 80
chromatid, 52, 65
chromosomal abnormalities, 16, 19, 20,
    109, 114, 123, 164
chromosomal alterations, 25, 48, 62, 146
chromosomal anomalies, vii, 1, 3, 47, 148
chromosomal instability, 47, 63, 124
chromosome, vii, 2, 3, 13, 14, 16, 17, 18,
    19, 20, 21, 25, 48, 49, 50, 55, 56, 64, 65,
    67, 70, 80, 103, 106, 107, 108, 109, 112,
    113, 122, 123, 125, 131, 133, 145, 153,
    155, 164
chronic lymphocytic leukemia, 70, 132, 135
chronic myelogenous, 13, 109, 110
chronic obstructive pulmonary disease, 140
cigarette smoking, 124
circulation, 70, 77, 162
cirrhosis, 138, 168
City, 156
classes, 46, 48, 62
classification, 5, 8, 11, 18, 108, 115
cleavage, 31, 32, 69
CLIA, 61
clinical application, 98, 165
clinical diagnosis, viii, 29, 65, 78, 143
clinical examination, 158
clinical oncology, 62
clinical symptoms, 80
clinical trials, 75

clone, 3, 25, 35, 36, 37, 120
cloning, 30, 31, 42, 69
clustering, 31
clusters, 31, 123
coding, viii, 33, 34, 66, 69, 144, 148, 152,
    156
codon, 148
coenzyme, 151
cognitive function, 152
collaboration, 61, 165
collagen, 74, 134
colon, 6, 11, 39, 54, 68, 69, 76
colon cancer, 69, 76
color, 16, 17, 20, 25, 40, 105, 107, 108, 130,
    145
colorectal adenocarcinoma, 110
colorectal cancer, 24, 26, 54, 65, 67, 69, 71,
    76, 114, 116, 120, 122, 127, 130, 131,
    136, 139
commercial, 158
community(ies), 32, 151, 152, 162
comparative analysis, 104
compilation, vii
complement, 48, 56, 65, 78
complementarity, 31
complementary DNA, 29, 55
complexity, 17, 23, 41, 47, 48, 49, 53, 63
composition, 32, 123
compounds, 50
computer, 12, 18
condensation, 5, 14
confidentiality, 99, 162
conformity, 54
conjugation, 11
connective tissue, 6
consensus, 49, 79, 97, 139
consent, 161
construction, 35, 105
control group, 71
corpus callosum, 151, 168
correlation(s), 48, 51, 66, 71, 110, 140
cost, 22, 26, 31, 37, 79, 99, 149, 154, 156,
    157, 158, 159, 160, 161, 162
counseling, 154, 157, 168
covering, 6, 20

CSCs, 73
CSF, 45, 94
CT, 11, 124
CT scan, 11
culture, 12, 13, 46
cure, 140
current limit, 26
CV, 127, 139
cycles, 27, 31, 32, 37, 53, 121
cyst, 77
cystic fibrosis, 146, 148, 149, 155, 166
cytochrome, 168
cytogenetics, 4, 13, 14, 16, 19, 21, 24, 40,
    49, 63, 104, 105, 108, 111, 112
cytokines, 79
cytokinesis, 4, 56
cytology, 10, 16, 106
cytometry, 64, 130
cytosine, 34, 54
cytoskeleton, 57

D

data analysis, 22, 115
data set, 24, 42, 160
database, 49, 130, 153, 162
deaths, 5
defects, 24, 47, 51, 55, 56, 154
defence, 56
deficiency, 151, 168
deficit, 51, 146, 166
degenerate, 32, 108
degradation, 57, 74, 129
deltaF508, 166
denaturation, 31
Denmark, 130
dephosphorylation, 129
depth, 164
deregulation, 59
derivatives, 25
desorption, 38
destruction, 39
detectable, 26, 36, 60, 71
detection system, 37
deviation, 44

diabetes, 139
diagnostic lab, viii, 19, 61, 80, 144, 145,
    146, 152, 155, 156, 157, 159
differential diagnosis, 150, 159
digestion, 22, 23, 26, 36, 92
dilation, 158
diploid, 56, 63, 65
disability, 145, 146, 148, 153, 154, 155,
    157, 165, 166, 171
discrimination, viii, 144, 147, 163
disease gene, 161
disease progression, 71, 75
diseases, viii, 59, 70, 78, 98, 103, 135, 143,
    145, 146, 147, 148, 149, 150, 151, 154,
    155, 156, 157, 158, 160, 161, 162, 164,
    165, 166, 167, 168, 169
disequilibrium, 69
disorder, 19, 146, 149, 152, 154, 155, 163,
    164, 166, 169
distribution, 2, 150, 153, 167
diversity, 119, 123
DNA damage, 50, 58, 62, 124, 127, 128
DNA ploidy, 129
DNA polymerase, 27, 29, 31, 37, 38
DNA repair, 47, 48, 51, 54, 55, 58, 60, 63,
    127
DNA sequencing, 28, 29, 30, 64, 117, 118,
    120, 144, 145, 147, 164, 165, 171
DNA testing, 159
docetaxel, 97
doctors, 11
dosage, 148
down-regulation, 132
draft, 127, 145
Drosophila, 132
drug metabolism, 97
drug resistance, 75
drugs, 58, 62, 66, 97, 129
dyes, 12, 18, 20, 27, 29, 32, 42, 145
dysplasia, 72, 75

E

Eastern Europe, 149
editors, 103

electric field, 29
electron, 39
electrophoresis, 27, 29, 69, 117, 147
ELISA, 64, 93, 94, 95
elucidation, 109
employment, 163
encephalitis, 151, 152, 169
encoding, 45, 153, 168, 169
endocrine, 67, 101
endometrial carcinoma, 112
endonuclease, 49
energy, 165
environment, 5, 40
enzymatic activity, 47
enzyme(s), 4, 19, 26, 27, 42, 54, 56, 57, 58,
    59, 65, 79, 104, 129
epigenetic alterations, 66
epigenetic modification, 53
epigenetics, 126
epilepsy, 155, 170
epithelia, 70
epithelial cells, 59, 129
epithelial ovarian cancer, 71, 75, 114
epithelium, 6, 72, 75, 101
epitopes, 140
Epstein Barr, 73
Epstein-Barr virus, 138
equipment, 24, 42
erosion, 55
ESI, 39
ESO, 140
esophageal cancer, 136
esophagus, 136
estrogen, 131, 133
ethanol, 15
ethical issues, vii, 162, 164
ethnic background, 149
ethnic groups, 149, 151, 162
ethnicity, 149
etiology, 146
euchromatin, 52
eukaryotic cell, 4
euploid, 3
evidence, 5, 8, 65, 70, 157
evolution, vii, 106, 117, 124

excision, 10, 47, 51, 55
exclusion, 169
execution, 56
exome, viii, 34, 144, 145, 151, 157, 158,
    161, 162, 163, 167, 171
exons, 126, 148, 153, 156, 158
exonuclease, 152, 169
exposure, 118
extraction, 38
EZH2, 54, 95, 100, 126

F

fallopian tubes, 75
false positive, 67, 163
familial hypercholesterolemia, 152
families, 126, 148, 153, 157, 166
family history, 67, 149, 151
fat, 6, 132
FDA, 61
fertility, 164
FHIT gene, 124
fiber, 107
fibroblasts, 13, 34
fibrous tissue, 6
fidelity, 55
filtration, 21
fingerprints, 36
Finland, 130
first generation, 145
fixation, 104
flight, 38
fluid, 12, 74, 78, 79, 93, 140, 152
fluorescence, 13, 14, 15, 17, 18, 19, 20, 25,
    27, 29, 37, 38, 42, 104, 105, 106, 107,
    108, 109
follicle, 76
force, 5, 53
formation, 52, 68, 115, 124
founder effect, 150, 151, 159
fragile site, 50, 124, 132
fragments, 22, 23, 27, 29, 31, 32, 33, 34, 40,
    42, 100, 145
France, 103, 150, 153
free radicals, 55

functional approach, 21
funds, 130
fusion, 21, 28, 34, 52, 109, 124, 125

G

gastrointestinal tract, 6, 75
gel, 27, 29, 117, 147
gene amplification, 18, 50, 65, 114, 125
gene expression, 3, 34, 52, 54, 59, 60, 68,
    114, 115, 141
gene pool, 150
gene silencing, 52, 54, 126
genetic alteration, 15, 18, 58
genetic counselling, 149, 155
genetic disease, 146, 150, 151, 164, 167,
    171
genetic disorders, 146, 164
genetic drift, 150
genetic factors, 146
genetic information, vii, 145, 162
genetic mutations, 147, 148
genetic predisposition, 97
genetic screening, 147, 148, 151, 152, 164
genetic testing, viii, 2, 145, 150, 151, 153,
    154, 158, 159, 164
genetics, 8, 14, 22, 108, 122, 130, 146, 151,
    159, 163, 165, 167, 168, 170
genital warts, 73
genomic instability, 53, 54, 110, 122
genomic regions, 48, 132
genomics, 39, 61, 98, 99, 122, 147, 156,
    163, 165
genotype, 69, 110
genotyping, 23, 64, 69, 113, 117, 132, 171
gland, 6
glioblastoma, 73
glioma, 66
glucose, 56, 77
glutamate, 166
glutamine, 148
glycol, 21
glycolysis, 56
glycosylation, 2
grading, 67

growth, 3, 44, 51, 57, 58, 67, 74, 76, 94,
    100, 102, 103, 113, 134, 155, 156
growth arrest, 113
growth factor, 44, 67, 74, 94, 100, 102, 103,
    134
growth hormone, 44
guidelines, 10, 99, 139, 163

H

half-life, 138
haploid, 68
haplotypes, 28
HDAC, 59
head and neck cancer, 130
healing, 55
health, vii, viii, 1, 9, 61, 144, 147, 156, 157,
    162, 163
health care, 156, 162
health care system, 156
health insurance, 163
heat shock protein, 140
hematopoietic stem cells, 51
hepatitis, 70, 75, 76, 137
hepatitis a, 76
hepatocellular cancer, 70, 134
hepatocellular carcinoma, 70, 75, 124, 133,
    134, 137, 138
heterochromatin, 2, 3, 14, 50, 52, 103
heterogeneity, 3, 20, 47, 48, 52, 67, 69, 79,
    80, 125, 146, 153, 157
histidine, 51, 91, 100, 124
histochemistry, 64
histone(s), 2, 5, 34, 53, 54, 58, 66, 119, 126,
    127, 128
histone deacetylase, 128
history, 14, 104, 145, 153, 167
HIV, 73
HM, 106, 121, 123, 135
homeostasis, 4
homocysteine, 80
homologous chromosomes, 4, 18
hormone(s), 44, 45, 76, 79
hospitalization, 149
host, 30, 42, 73, 74, 98

hot spots, 49, 169
human body, 38, 145
human brain, 129
human chorionic gonadotropin, 139
human genome, vii, viii, 5, 9, 20, 22, 23, 26,
  29, 30, 36, 49, 53, 69, 108, 113, 115,
  118, 119, 120, 143, 145, 146, 150
human health, 145
human papilloma virus (HPV), 73, 137
Hungary, 153
Hunter, 121
hybrid, 20
hybridization, 14, 17, 18, 23, 24, 25, 26, 28,
  34, 42, 69, 91, 103, 104, 109, 110, 111,
  117, 125, 146, 155
hybridoma, 139
hydrogen, 38
hyperactivity, 166
hypermethylation, 54, 65, 126, 130
hypothesis, 77
hypothyroidism, 151
hypoxia, 55, 127

I

icterus, 76
ideal, 57, 69
identification, 11, 17, 18, 19, 38, 46, 50, 61,
  67, 69, 70, 76, 108, 109, 125, 145, 146,
  148, 151, 157, 161, 165, 167
identity, 162
IFN, 96, 152
IL-8, 94
image(s), 12, 17, 18, 20
imbalances, 19, 48, 111
immigrants, 150
immigration, 150
immune response, 78, 137, 140
immune system, 60, 72, 77
immunity, 73, 140
immunodeficiency, 54
immunohistochemistry, 19, 104
immunoprecipitation, 34
improvements, 37, 66, 164

in situ hybridization, 15, 16, 17, 19, 105,
  106, 107, 109
in vitro, 30, 44, 115, 124, 164
in vivo, 29, 42, 115, 128
incidence, 103, 147, 149, 152, 153, 154
India, 109
individualization, 98
individuals, 23, 98, 145, 150, 151, 152, 153,
  157, 162, 163, 171
induced bias, 118
induction, 77, 113, 138
induction chemotherapy, 138
industry, 98
infancy, 152, 154, 168
infant mortality, 147
infants, 152, 168
infection, 6
infertility, 51
inflammatory disease, 76
inheritance, 149, 164, 169
inherited disorder, 154
inhibition, 57, 59
inhibitor, 13, 80, 94, 99, 102, 128
initiation, 54, 61, 74, 131, 152, 158
injury, 51
institutions, 60, 61, 79
insulin, 94
integration, 50, 73, 100, 124
integrity, 53, 55
intellectual disabilities, 146, 158
intensive care unit, ix, 144, 164, 171
interface, 18
interferon, 133, 134, 169
International Classification of Diseases, 5
interphase, 15, 16, 17, 48, 49, 63, 105, 111,
  124, 125
intervention, 68, 157
intron(s), 33, 99, 156
inversion, 4, 18, 36, 63
ionizing radiation, 50
ions, 38
islands, 27, 50
isolation, 25, 38
isomerization, 129
issues, 98, 99

Italy, 153

J

Japan, 153
Jews, 149
jumping, 49, 55, 124

K

karyotype, 13, 14, 24, 48, 63
karyotyping, 15, 17, 18, 20, 24, 25, 42, 48,
   104, 108, 115, 116, 123, 124, 125, 146,
   163
keratinocyte, 113
kidney, 5, 51, 65, 68, 72, 80
kinetochore, 56

L

laboratory tests, viii, 2
landscape(s), 119, 120
laparoscope, 11
large intestine, 11
LDL, 80
lead, 4, 11, 18, 23, 41, 44, 48, 53, 54, 55,
   63, 99, 146, 153, 158, 163, 165
lesions, 24, 51, 57, 131
leukemia, 7, 13, 34, 49, 51, 68, 70, 77, 100,
   101, 102, 103, 109, 110, 119
LFA, 44
life sciences, 156
ligand, 135
light, 12, 18, 30, 39, 50, 93
liquid chromatography, 28, 69, 117
liver, 5, 13, 39, 51, 68, 69, 73, 74, 76, 77,
   132
liver cancer, 5
liver cirrhosis, 75
liver disease, 69, 74, 76
localization, 59, 109
loci, 17, 22, 25, 28, 67, 132, 155, 164
locus, 17, 21, 26, 40, 41, 112, 169
longevity, vii, 1, 52

luciferase, 30
luciferin, 30
lung cancer, 5, 65, 71, 76, 106, 118, 126,
   130, 135, 136, 137, 141
lung metastases, 106
Luo, 135
lymph, 7, 8, 10, 83
lymph node, 7, 8, 10
lymphatic system, 7
lymphocytes, 13, 44, 80
lymphocytosis, 152
lymphoid, 7, 101
lymphoma, 7, 66, 70, 73, 83, 99, 105, 124,
   135
lysine, 2, 54, 58

M

mAb, 39
machinery, 35
macromolecules, vii, 1, 47, 74
magnet, 26, 34
magnetic field, 12
majority, 62, 68, 70, 146, 147, 153, 155,
   157
MALDI, 38
malignancy, 7, 19, 59, 70, 97, 110, 127
malignant cells, 65, 77
malignant melanoma, 24, 114, 116
malignant tumors, 11, 47, 57, 65, 138
mammalian cells, 122
mammals, 74
man, 13
management, vii, 1, 5, 9, 44, 79, 146, 154,
   167
manipulation, 57
mantle, 104
mapping, 17, 26, 36, 42, 51, 107, 119
marketing, 61
marrow, 6, 15, 92
mass, 38, 140
mass spectrometry, 38, 140
matrix, 21, 29, 38
matter, 123, 170
MB, 50, 111, 153

MBP, 96
measurement(s), 75, 77, 111, 139
media, 13
medical, vii, 1, 28, 44, 97, 145, 149, 157,
    158, 161, 163, 167
medicine, viii, 12, 38, 40, 73, 80, 97, 139,
    144, 157, 161, 165, 168
Mediterranean, 149, 167
medulloblastoma, 26, 73, 116
meiosis, 4
melanoma, 24, 26, 68, 73, 106, 115
mellitus, 139
membranes, 6
menstruation, 75
mental retardation, 152, 155, 166, 169, 170
mesothelioma, 6
messengers, 45
Metabolic, 56, 147, 170
metabolic disorder(s), 44, 155, 159
metabolism, 3, 56, 127, 132, 151, 154, 169,
    170
metabolites, 76, 139
metalloproteinase, 102
metaphase, 12, 13, 14, 15, 16, 17, 18, 20,
    41, 48, 50, 54, 55, 63, 80, 103
metaphase plate, 55
metastasis, 3, 8, 21, 54, 75, 77, 133, 134
methanol, 13
methyl group(s), 54
methylation, 5, 27, 34, 48, 53, 54, 60, 65,
    92, 93, 94, 96, 116, 126, 127, 130, 135
microarray technology, 24, 113
microcephaly, 152, 155
microenvironments, 78
microRNA, 132, 133, 134, 135, 136
microsatellites, 166
microscope, 10, 14, 17, 19
microscopy, 15, 17, 18, 104, 108
migration, 56, 58, 133, 134, 150
mimicry, 115
mission, 162, 163
mitochondria, 127
mitochondrial DNA, 4, 56, 60, 66, 132
mitochondrion, vii
mitosis, 12, 13, 56, 63, 129

modelling, 40
models, 55
modifications, 5, 38, 54, 59, 80, 119, 126,
    127
modifier gene, 126
molecular biology, 9, 62, 104, 107, 144
molecular cytogenetics, 104, 106
molecular diagnosis, viii, 144, 146, 154,
    157, 161, 162, 163
molecular medicine, 105
molecules, 2, 28, 30, 31, 37, 38, 45, 58, 120,
    144
monoclonal antibody, 75
monosomy, 21
morbidity, 5
Morocco, 153
morphology, 19
mortality, 5, 103
mosaic, 21, 151
motif, 49
MR, 104, 107, 108, 115, 116, 120, 122, 123,
    131, 132
MRI, 11
mRNA(s), 2457, 67, 131, 133
mtDNA, 68
mucosa, 6
mucus, 6
multicellular organisms, 128
multiple myeloma, 50, 105
multiplication, 58, 157
muscles, 6
muscular dystrophy, 148, 166
mutant, 3, 58, 68, 149, 151, 152, 168
mutation, viii, 2, 4, 47, 53, 54, 64, 66, 69,
    82, 83, 92, 118, 126, 131, 144, 147, 148,
    152, 153, 154, 155, 157, 159, 160, 163,
    165, 166
mutation rate, 53, 126, 155, 165
myelodysplastic syndromes, 54, 115, 126

N

Na$^+$, 122
nanometers, 37
nasopharyngeal carcinoma, 73

neoplasm, 20, 52, 67, 101
nerve, 44
nervous system, 170
nested PCR, 27
Netherlands, 153
neuroblastoma, 27, 75, 101, 116, 132, 168
neurodegeneration, 51
neurologist, 158
neuropathy, 151
neutral, 23
nevus, 106
next generation, viii, 33, 34, 118, 143, 145, 156
nodal involvement, 8
nodes, 10
nodules, 77
North America, 151, 168
nuclei, 13, 16, 17, 49
nucleic acid, 14, 29, 37, 48, 73, 116, 164, 166, 171
nucleolus, 57
nucleoprotein, 53
nucleosome, 2, 128
nucleotide sequence, 3, 33
nucleotides, 22, 28, 30, 31, 37, 38, 69
nucleus, vii, 19, 52, 57, 124, 125

O

obesity, 155
oesophageal, 68, 72
OH, 27, 57
oligonucleotide arrays, 22, 23, 115
oncogenes, 46, 51, 54, 65
oncogenesis, 50, 59, 69
opportunities, 151, 156
optimization, 66
organ(s), 6, 11, 12, 13, 51, 76, 79, 137, 154
organism, 144
osteoporosis, 51
outpatient, 10
ovarian cancer, 11, 24, 71, 75, 76, 112, 114, 136, 138, 140
ovarian tumor, 75, 112
overlap, 16, 48, 152, 155

oxidative stress, 56
oxygen, 77

P

p53, 58, 68, 91, 92, 93, 94, 113, 128, 140
Pacific, 37, 137
palate, 101
palliative, 44
pancreas, 39, 52, 76, 137
pancreatic cancer, 24, 66, 75, 76, 114, 130, 134
paradigm shift, viii, 144
parallel, vii, viii, 1, 24, 28, 29, 38, 40, 47, 48, 63, 117, 118, 119, 143, 145, 156, 164
parents, 145, 164
Parnes, 138
participants, 162, 163
pathobiology, vii, viii, 2, 97
pathogenesis, 60, 98, 169
pathologist, 10
pathology, 10, 61, 137
pathophysiological, 69
pathways, vii, 1, 38, 40, 48, 55, 56, 59, 60, 63, 78, 128, 170
patient care, 66, 99, 147, 163
PCR, 20, 22, 23, 24, 25, 27, 28, 30, 31, 32, 64, 71, 82, 83, 91, 92, 93, 94, 95, 96, 97, 101, 104, 108, 116, 117, 136, 147, 148, 166
peptide, 59, 95
peripheral blood, 44, 54, 71, 73, 137
peripheral neuropathy, 168
permit, 52
PET scan, 12
PGD, 164, 166
pH, 38, 77
phage, 35
pharmacogenomics, 97, 157
pharmacological treatment, 158
phenotype(s), 47, 48, 55, 59, 110, 131, 149, 151, 154, 155, 159, 164, 170
phenylketonuria, 151
Philadelphia, 13, 80, 109, 166
phosphorylation, 53, 57, 59

physicians, 154
PI3K, 57, 134
placenta, 76
plants, 29
plaque, 6
plasma cells, 6
plasma membrane, 45
plasmid, 25
platform, 34, 37, 39, 43, 98, 99, 132, 136, 156
platinum, 71
ploidy, 20, 48, 56, 63, 65
PM, 134, 168
point defects, 56
point mutation, 33, 66, 146, 153, 163
polar, 56, 147
policy, 162, 165, 167
polyacrylamide, 29
polymerase, 27, 31, 37, 71, 120, 147, 148
polymerase chain reaction, 71, 147, 148
polymorphism(s), 43, 49, 65, 70, 97, 113, 114, 120, 130, 132, 133, 145, 147, 153, 160, 164
polypeptide, 101, 102
pools, 150
population, 17, 26, 34, 63, 69, 72, 73, 98, 138, 148, 149, 150, 151, 152, 155, 157, 159, 160, 167, 169
population structure, 150, 167
portal vein, 133
predictability, 98
pregnancy, 75, 152
preparation, 12, 37
prevention, 58
primary tumor, 52
prior knowledge, 19, 25, 80
probands, 145
probe, 15, 16, 17, 20, 24, 28, 32, 34, 115, 116, 148, 166
professionals, 163
prognosis, 47, 56, 62, 68, 70, 73, 75, 76, 77, 79, 99, 134, 135, 136, 137
project, 51, 69, 145, 158
prolactin, 94

proliferation, 3, 5, 44, 57, 59, 67, 77, 129, 134, 137, 159
proline, 59, 129
promoter, 54, 65, 126, 130
prophase, 14, 80
prophylactic, 137, 158
prostate cancer, 11, 24, 50, 53, 56, 67, 73, 74, 80, 100, 114, 122, 124, 136, 138
prostate carcinoma, 65, 77
prostate specific antigen, 11
prostrate cancer, 5
protection, 64, 75, 163
protein components, 57
protein kinases, 44
protein-protein interactions, 59
proteins, 2, 3, 38, 39, 45, 46, 54, 55, 57, 58, 59, 60, 61, 74, 75, 77, 78, 79, 93, 100, 122, 131, 140, 144, 145
proteolysis, 74, 129
proteome, 68, 74
proteomics, 38, 39, 61, 68, 98, 120, 140, 156
proto-oncogene, 101
pseudogene, 99
psychiatric disorders, 145, 163, 165
PTEN, 57, 101, 134
public health, 151
purification, 117
pyrophosphate, 30, 42

Q

quality of life, vii, 1
quantification, 16, 73, 98, 110, 116, 148, 166
query, 10
quinacrine, 13

R

race, 120
radiation, 5, 20, 55, 58
radio, 12, 39
radioactive tracer, 12

RARB, 92, 93, 96, 102
RASSF1, 93, 102
RB1, 67
reactions, 28, 29, 42
reading, 3, 144
reagents, 24, 42
real time, 37, 38
reality, 40, 138, 156, 159
receptors, 58
reciprocal translocation, 25, 41, 50
recognition, 26, 55, 77, 102, 154
recombination, 2, 4, 49, 50, 53, 55, 123
recommendations, 162, 170, 171
reconstruction, 17, 40
recurrence, 77
red blood cells, 7, 13
redundancy, 35, 36, 69
regulatory bodies, 61
relatives, 162
remodelling, 59, 116, 125
renal cell carcinoma, 5, 72, 125
repair, 4, 47, 51, 55, 57, 58, 63, 65, 101,
    122, 127, 130, 136, 152
replication, 4, 50, 55, 131
reporters, 140
repression, 54
repressor, 113
reproduction, 167
researchers, viii, 1, 71, 146, 162, 163
residual disease, 15, 110
residues, 34, 54
resistance, 67, 98, 116, 133, 135
resolution, viii, 14, 17, 20, 22, 23, 25, 33,
    34, 35, 36, 38, 40, 41, 42, 43, 52, 107,
    110, 111, 113, 114, 117, 119, 123, 125,
    143, 146
resources, 60, 115
response, 6, 56, 58, 60, 61, 62, 63, 67, 68,
    71, 73, 74, 75, 77, 93, 122, 130, 134,
    136, 157
responsiveness, 66, 67
restoration, 21
restriction enzyme, 22, 26, 36, 42
restriction fragment length polymorphis, 23,
    37

retardation, 156
retinitis, 157
retinitis pigmentosa, 157
retinoblastoma, 67, 130
retroviruses, 73
reverse transcriptase, 57, 102
ribosome, 57
ring chromosome, 18, 48, 52
rings, 63
risk(s), 51, 62, 65, 67, 68, 69, 70, 75, 78, 80,
    107, 130, 131, 133, 139, 149, 151, 157,
    160, 161, 165
risk assessment, 68
RNA(s), 2, 26, 30, 33, 34, 38, 47, 57, 60,
    61, 64, 66, 67, 69, 70, 72, 78, 92, 93, 94,
    95, 96, 98, 119, 121, 132, 134, 147
roots, 151
Rouleau, v

S

safety, 60
saliva, 10, 78
scatter, 39
scattering, 39
schizophrenia, 145, 146, 155, 165, 166, 170
science, 12, 98, 145
scientific knowledge, 163
sclerosis, 102
scoliosis, 155
scope, 62
secretion, 6, 74
segregation, 4, 47, 56, 102
selectivity, 60
semiconductor, 38
senescence, 21, 51, 112
sensitivity, 20, 22, 66, 71, 75, 78, 98, 104,
    134, 147, 155, 160, 163
Serbia, 153
serine, 57, 59, 74
serum, 38, 60, 65, 69, 73, 74, 75, 76, 77, 78,
    79, 94, 98, 121, 130, 135, 136, 138, 139,
    171
services, 155, 159
sex, 109, 149, 164

shape, 39
shock, 77, 140
showing, 106, 157
siblings, 153
sickle cell, 149, 167
sickle cell anemia, 149
side chain, 58
signal transduction, 54
signalling, 40, 56, 57, 59, 67, 78
signals, 17, 19, 23, 25, 45, 58
signs, 10
single test, vii, 1
siRNA, 67
skeletal muscle, 6
skin, 6, 13, 33, 73, 147
skin cancer, 73
smoking, 5, 139
smooth muscle, 6, 101
SNP, 23, 24, 25, 42, 64, 69, 113, 114, 115,
   131, 132, 146, 164, 171
social context, 150
sodium, 15
soft tissue sarcomas, 56
software, 20, 31
solid phase, 34
solid tumors, 33, 56, 58, 80, 110, 133
solution, 12, 13, 15, 34
somatic cell, 47
somatic mutations, 33, 68, 118
Spain, 153
spastic, 151, 153, 158, 168, 169
spasticity, 158
species, 3, 29, 134
spectrophotometry, 95
speculation, 69
spinal cord, 152
spindle, 55, 63, 80
sputum, 10, 38, 65, 79
squamous cell carcinoma, 6, 52, 65, 72, 78,
   102, 125, 136
stability, 34, 59, 69
stakeholders, 79, 99
state(s), 2, 4, 23, 44, 50, 60, 61, 79, 128,
   150, 154, 164
statistics, 24

stem cells, 34, 57, 81, 137
stimulation, 21, 58, 77
stomach, 7, 12, 76
storage, 98, 154
stratification, 150, 168
stress, 50, 77, 128
stress response, 128
structural changes, 62
structural characteristics, 17
structural protein, 40
structural variation, 63, 119
structure, 2, 4, 33, 34, 38, 47, 59, 60, 120,
   123, 138, 145, 151
subgroups, 24
substitution, 32
substrate(s), 19, 20, 38, 58
Sun, 135
supervision, 167
suppression, 21, 51, 58, 60, 113, 128, 135
surveillance, 76
survival, 5, 16, 56, 63, 66, 71, 72, 73, 114,
   129, 133, 136, 137
susceptibility, 67, 122, 130, 131, 163
Sweden, 130
symptoms, 5, 152, 155
syndrome, 54, 109, 126, 148, 151, 152, 155,
   168, 169, 170, 171
synthesis, 29, 30, 31, 37, 38, 76, 120

T

T cell, 72, 83, 137
tandem repeats, 64
target, 15, 16, 17, 19, 21, 22, 39, 54, 57, 58,
   60, 63, 71, 73, 108, 128, 129, 148, 163
Tay-Sachs disease, 149, 167
T-cell receptor, 102
TCR, 51, 82, 102
teams, 60
techniques, 9, 11, 12, 13, 14, 15, 19, 38, 40,
   62, 63, 64, 65, 68, 80, 103, 105, 117,
   145, 147, 164
technological advances, vii, 1, 9, 47, 114
technological progress, viii, 2, 146

technology(ies), viii, 5, 8, 12, 24, 28, 29, 30, 31, 32, 37, 38, 39, 40, 60, 78, 79, 98, 105, 107, 113, 116, 117, 119, 139, 143, 145, 147, 156, 161, 171
telomere, 53, 55, 57, 125
telomere shortening, 53
testing, viii, 98, 99, 111, 143, 148, 149, 150, 154, 155, 156, 159, 160, 166, 170, 171
TGF, 44, 67, 95, 102
thalassemia, 149, 167
therapeutics, 57, 61, 62, 66, 67, 128
therapy, 39, 40, 51, 56, 61, 62, 66, 72, 75, 77, 79, 98, 116, 120, 126, 157
threonine, 45, 57, 59, 102
thrombus, 133
thyroglobulin, 76, 77, 139
thyroid, 68, 76, 77, 131, 139, 140
thyroid cancer, 68, 76, 77, 139
thyroid gland, 76
thyrotropin, 139
thyroxin, 76
tissue, 3, 5, 6, 7, 8, 10, 11, 12, 15, 19, 24, 38, 39, 52, 57, 60, 69, 74, 77, 79, 80, 98, 104, 109, 120, 148
TNF, 135
tobacco, 118
total cholesterol, 80
total internal reflection, 37
toxicity, 61, 62
TP53, 82, 102
TPA, 139
traits, 149
transcription, 2, 3, 34, 50, 68, 100, 102, 128, 140
transcription factors, 128
transcriptomics, 119
transcripts, 34, 109
transfection, 21, 102
transformation, 4, 24, 44, 47, 56, 58, 61, 65, 73, 78, 115, 128, 129
transforming growth factor, 44
translation, viii, 2, 3, 57, 68, 99, 152
translocation, 13, 16, 25, 36, 40, 49, 50, 54, 63, 80, 82, 83, 102, 105, 122, 123, 124, 164

transmission, 164, 167
transport, 98
treatment, vii, viii, 1, 5, 8, 9, 14, 39, 44, 59, 62, 63, 70, 75, 77, 79, 80, 98, 99, 104, 140, 143, 149, 154, 157, 158, 162, 168
trial, 66, 72
triglycerides, 80
triiodothyronine, 76
triploid, 56
trisomy, 21, 171
trypsin, 13, 94
TSH, 77
tumor cells, 15, 17, 26, 27, 52, 54, 56, 59, 66, 73, 78, 131, 136
tumor development, 4, 50, 58
tumor growth, 57, 60, 134
tumor progression, 58, 77, 78, 110, 127
tumorigenesis, 110, 112, 122, 132, 140
tumors, 7, 11, 18, 25, 27, 30, 40, 48, 49, 50, 54, 55, 56, 57, 58, 59, 61, 63, 65, 66, 72, 73, 76, 77, 78, 80, 98, 110, 122, 123, 126, 129, 132, 137, 140
Turkey, 153
twins, 130
tyrosine, 45, 58, 99, 101, 121, 128

U

ubiquitin, 129
ultrasound, 12
umbilical cord, 13
unions, 150
United Kingdom (UK), 137, 153
United States (USA), 5, 103, 104, 107, 110, 115, 116, 117, 120, 123, 125, 127, 128, 149, 163
urinary bladder cancer, 67, 131
urinary tract, 76
urine, 10, 38, 75, 76, 79, 80, 106, 151, 168
uterus, 76

V

vaccine, 137

vagina, 76
validation, 61, 66, 72, 79, 98
variations, 22, 23, 34, 78, 145, 157, 171
vein, 12
Vermeer, 169
viral infection, 5
virus infection, 137
viruses, 46, 50, 73
visualization, 15, 19, 107

WHO, 5, 103
wild type, 58
Wnt signaling, 131
workers, 29, 51
workflow, 32
worldwide, vii, 1, 103, 151, 153, 154

W

wavelengths, 38
web, 162
wells, 30
western blot, 96
white blood cells, 6, 13
white matter, 152

X

X chromosome, 155, 169

Y

yeast, 119
yield, 13, 24, 145, 146, 159, 160, 162, 163
young adults, 7